Computer Science, Technology and Applications

Advances of Machine Learning in Clean Energy and the Transportation Industry

Computer Science, Technology and Applications

Additional books and e-books in this series can be found on Nova's website under the Series tab.

COMPUTER SCIENCE, TECHNOLOGY AND APPLICATIONS

ADVANCES OF MACHINE LEARNING IN CLEAN ENERGY AND THE TRANSPORTATION INDUSTRY

PANDIAN VASANT
VALERIY KHARCHENKO
VLADIMIR PANCHENKO
J. JOSHUA THOMAS
AND
GERHARD-WILHELM WEBER
EDITORS

nova
science publishers
New York

DOI: https://doi.org/10.52305/SJDR3905

NOTICE TO THE READER

Library of Congress Cataloging-in-Publication Data

ISBN: 978-1-68507-211-7

Published by Nova Science Publishers, Inc. † New York

CONTENTS

PREFACE

We as the editors have hoped that this book will become one of the most helpful and valuable reference books of 2021 in fields where *machine learning* serves for *clean energy* and *transportation industry*. This book intends and has succeeded in encompassing emerging research on new and modern, upcoming and future resources and means, algorithms and methodologies, in the emerging transportation sector and its fascinating related topics. Owed to this work's international aspects and directions, it has become a cutting-edge resource, outlining world-class results and achievements made in any continent by the scientific support of operational research, artificial intelligence, cognitive sciences, physics, the sectors or renewable energy, electrical engineering, electro-technologies, etc., and by our befriended research communities.

Our book project was accepted by the house of *Nova Scientific Publishers* for its comprehensive analytical and practical appeal and promise on a global stage. This handbook and compendium along with our investigations on renewable energy and power supply challenges are a considerate scholarly resource which examines the efficient use of modern electric resources and renewable natural sources. They should have a positive impact on sustainable development in our cities and rural countryside, in our oceans and the sky, thus, for mobility and communication, collaboration and peace in and among our countries, nations and our creatures in the world.

In recent decades, attention and interest to clean energy has been growing exponentially, along with a continuously increasing concern about our natural environment and our health, our natural resources and our atmosphere. This development has especially been due to a decrease in both the cost of both installed capacity of converters and the cost of generated energy. Such amazing successes could be achieved thanks to advancements and improvements of modern technologies for the production of converters, an enhancement in the efficiency of use of incoming energy, an optimization of the operation of converters and analysis of data obtained during the operation of systems with the possibility of production planning. In the transportation industry where technologies have also been improved from year to year, the use of clean energy plays an important role as well. The transportation industry is growing, and machinery and systems are becoming more autonomous and robotic, which is no longer possible without intelligent computing and operational research (OR), machine learning and artificial intelligence (AI), optimization and optimal control theories, statistics and computer science, planning and working with a huge amount of "ocean of data".

This work and research compendium reflects the latest, the emerging and also future research on the field of machine learning, OR and AI, discussing real-world application problems associated with new innovative renewable energy methodologies as well as cutting edge technologies in the quickly expanding transport industry. The requirements and demands of problem solving have grown with exponential progression, and new machine learning, OR and AI algorithms, methods and techniques have reduced the scope of data coverage worldwide. Recent progress in data technology (DT) have contributed to a reduction of gaps within antagonisms and trade-offs, and in the coverage of domains around the globe.

This book includes newest advances in theory, methods and practical implementations with machine learning, quantum optimization, modern technology in transport industry, and innovative technologies in transport education, systems based on renewable energy conversion, business process models and applications in renewable energy, clean energy, and climate change. Special topics on energy resources and renewable resources, hydropower energy and wind power, solar power and geothermal energy, tidal and wave power, biomass and biofuel, hydrogen and electrical vehicles, with artificial intelligence and operational research, optimization and control, neuroscience and soft computing, have been welcome also.

PART A: CLEAN ENERGY

In *Chapter 1* called as "RES-based multipurpose plant for hydrogen production", V. Adomavicius presents a concept of novel multipurpose plant for hydrogen production based on the presently most appealing solar and wind power. Solar and wind power sources are chosen for hydrogen production as well since they successfully compete with other electricity production technologies and their future is even more promising. The concept of the proposed plant for hydrogen production is explained by a block diagram of the plant's structural units and simplified diagram of the control system of technological processes taking place in this plant. Possible and potentially bountiful targets for the use of the RES-based power plant are also described.

In *Chapter 2* dedicted to "Developing A Bayesian Network To Model Environmental, Organizational, And Human Risk Factors: A Case Study On Wind Turbines", Maryam Ashrafi notices that technological systems embedded in human, organizational and environmental contexts reveal various interactions. Most risk assessment techniques encounter limitations. She proposes an integrated risk assessment framework for technological systems using a Bayesian network to consider various levels and their interactions. The proposed model is applied to assess risk in a wind turbine as an instance of complex power generating systems consisting of structural, electrical, and mechanical components interacting with human resource and organizational and environmental factors. The new methodology can be expanded to an entire wind farm or other complex technological systems.

In *Chapter 3* called “Digital technologies for the implementation of intelligent diagnostics of the isolation of power supply systems with isolated neutral in the operating mode”, Svetlana Ovchukova, Nadezhda Kondrateva, Andrey Shishov, Roman Bolshin and Maria Krasnolutskaya underline that 75% of emergencies at power supply facilities are caused by single-phase ground faults. Currently methods of thermal imaging and ultrasonic testing do not allow accurate identification of the cause of accidents in electric power distribution systems. The authors suggest digital technologies for intelligent diagnostics of the insulation of power supply system with insulated neutral in operating mode, allowing for on-line diagnostics and analyzing of data obtained for the early prevention of emergency situations. They particularly refer to smart agribusiness.

In *Chapter 4* dedicated to “Irrigation System Of Agricultural Fields With The Use Of Solar Energy”, Leonid Yurievich Yuferev and Alexander Sergeevich Parakhnich analyze the regions of the country where crops are being dyed in order to identify the possibility of using renewable sources to ensure the required irrigation rate. To calculate the power of the solar power plant, it was necessary to identify the power consumption of the electric pump. The standard method showed a large error. To improve the accuracy, a proprietary calculation method was developed based on operational tests of circulation pumps using machine learning through a regression algorithm. The new technique makes it possible to reduce the error from 20% to 2-3%. Furthermore, a schematic diagram of the operation of an autonomous irrigation system has been designed, and choice of equipment made.

In *Chapter 5* about “Hybrid simulation of strategies for the development of the renewable energy market in the regions”, P.N. Kuznetsov, D. Yu. Voronin, L.Yu Yuferev and V. P. Evstigneev address the constant need to find a compromise between electricity demand and its available volume at any time. Demand is characterized by high volatility, which significantly complicates the adoption of effective decisions on an optimal strategy for operational dispatch control of the power system and the organization of trade. Renewable energy sources need for effective means of dynamic correction of power energy production produced by traditional generation. The authors develop effective tools for countering the a priori uncertainty of demand and supply dynamics for the distributed electricity market through adaptive procedures to find compromise solutions for proactive management of the regional electricity market.

In *Chapter 6* devoted to "RES-based power plants versus polluting power plants: pros and cons", V. Adomavicius reviews the mainly used electric power generation technologies: 3 of them are based on renewable energy sources (solar, wind, and hydropower plants) and the remaing 3 (on the fossil fuels of natural gas, coal, as well as nuclear power plants). Disadvantages and advantages of all types of power plants are described. The information answers why RES-based power plants currently attain top positions in electricity generation worldwide and in lowest indicators of environmental pollution. The chapter concludes that this breakthrough in RES-based technologies should be successful while polluting power generation technologies will gradually be abandoned. There is a good chance that the green revolution in the global energy sector will prevail.

In *Chapter 7* called "A comprehensive study of system building blocks for Radio Frequency energy harvesting", Bhuvnesh Khantwal, Reeta Verma, and Paras state that our environment has been condensing by various Radio Frequencies (RF) waves by the rapid increase in wireless communication technologies over the last decades. Researchers have extended the idea of energy harvesting to include RF harvesting. This review paper presents a comprehensive study of the complete Radio Frequency Energy Harvesting (RFEH) system. Basic building blocks are antenna, matching circuit, rectifier and power management unit. Various circuit topologies involved in different blocks of the system are compared. Furthermore, the advancements and challenges in this modern energy harvesting field are also emphasized.

PART B: TRANSPORTATION INDUSTRY

In *Chapter 8* named as "The Management Of Community Participation In Rural Infrastructure Development In The Cuu Long River

Delta Area, Vietnam", Nguyen Xuan Quyet, Dinh Tuan Hai, Duc Anh Nguyen and Nguyen Viet Luan address the total length of rural roads in Vietnam, which has increased to 272,861 km, accounting for 82% of the length of the road network. A short-term and reasonable investment to the road network is estimated a cost of $6.6 billion. Government policies and management mechanisms have shown a lack of transparency here, leading to law abuse and misdemeanor in local project management. This study reviews the real situation of community participation in rural infrastructure development until 2020, and it proposes several approaches to improve the participation of community and its effectiveness in rural infrastructure development

In *Chapter 9* named as "Warning System for Cracked Pipes in Autonomous Vehicles", Yair Wiseman states that the looming autonomous vehicles will be driven by the vehicles themselves. Therefore, autonomous vehicles will be less inspected because human drivers often listen to the engine and its noises. The passengers in autonomous vehicles are usually bothered with other issues and sometimes even sleep. So there will be no effective human inspection on the engine and specifically on the pipe system. This paper offers an automatic system which warns about cracked pipes before they are out of use.

In *Chapter 10* on a "Contribution Of Machine Learning To Rail Transport Safety", Habib Hadj-Mabrouk notices that seriousness of accidents and incidents in rail transport have forced all players in the rail system to set up an "experience feedback" system (REX) to improve operational safety. Knowledge of accidents and incidents results essentially from the contribution of lessons learned and experiences acquired. This process can be facilitated by the use of AI and machine learning. The approach proposed is a hybrid method built around 3 learning algorithms and 1 knowledge-based system (KBS). This risk of accident requires the implementation of new prevention or protection measures during the various safety analyses of hardware and software equipment. CBR makes it possible to look for the most similar cases to this new risk of accident and proposes the measures.

In *Chapter 11* on "The power of variable freeing and variable sum bounds in solving the linear knapsack problem", Elias Munapo presents a technique of using variable freeing and variable sum bounds in solving the general linear knapsack problem. The lower and upper bounds of a sum of variables of any linear knapsack model is accurately determined. The basic variable in the continuous optimal solution of the knapsack problem, also known as the variable with the largest weight freed of integral restrictions before solved by the standard branch and bound algorithm. A linear knapsack problem with variable sum bounds and a freed basic variable is easier to resolve by the standard branch and bound algorithm. There are numerous important applications of the knapsack problem, especially in the transportation sector.

The conceptual and empirical chapters of this compendium stand for significant advancements with numerous research perspectives of analysis, adopted for unfolding and displaying selected challenges and themes, approaches and insights, methodologies and solutions. They design and foster innovative progress for the practice. Our book is pursuing a multidisciplinary, creative and holistic research view, in which diverse science and application fields can come together and mutually immerse, get involved and integrated for high purposes.

We *editors* hope that the chosen fields and picked subjects represent a core relevance and promise of international research on the emerging complex problems of *Machine Learning* within *Clean Energy* and *Transportation Industry* by the results and methods of *AI* and *OR*. We are very thankful to the house of *Nova Scientific Publishers* for the honor of accommodating this book as a pioneering endeavor in science, implementation and practice. Special gratitude is extended to the respected directors, editors and managers of *Nova Scientific Publishers* as well as to its *editorial management* and *staff*, for their approval, continuous interest and support in every regard. We thank all the authors for their hard work and readiness to share their newest deep understanding, their discoveries and results, their hope and cheer with our entire research community. We also express our thanks to the anonymous reviewers of the chapters submitted for their care- and helpful assessments. Now we very much hope that this book will stimulate collaboration and community of high quality, good and new life at a worldwide stage.

Book Editors,

Dr. Pandian Vasant
MERLIN Research Centre, Ton Duc Thang University, Vietnam
E-mail: pvasant@gmail.com

Prof. Valeriy Kharchenko
Federal Scientific Agroengineering Center VIM, Russia
E-mail: kharval@mail.ru

Dr. Vladimir Panchenko
Russian University of Transport, Russia
E-mail: pancheska@mail.ru

Dr. J. Joshua Thomas
UOW Malaysia; KDU Penang University College, Malaysia
E-mail: jjoshua@kdupg.edu.my

Prof. Gerhard-Wilhelm Weber
Poznań University of Technology, Poland
E-mail: gerhard.weber@put.poznan.pl

PART A: CLEAN ENERGY

In: Advances of Machine Learning … ISBN: 978-1-68507-211-7
Editors: Pandian Vasant et al.

Chapter 1

RES-BASED MULTIPURPOSE PLANT FOR HYDROGEN PRODUCTION

Vytautas Adomavicius*
Department of Electric Power Systems,
Kaunas University of Technology, Kaunas, Lithuania

ABSTRACT

The concept of a novel multipurpose plant for hydrogen production based on the most promising solar and wind power is presented in this chapter. Usually solar and wind power generating systems have only one purpose – to produce electric power and supply it to the main electric grid or to stand-alone power users. Hydrogen production in this multipurpose system is one of the most important goals, but not the only one. Solar and wind power sources are chosen for hydrogen production because presently they compete very successfully with other electricity production technologies and their future is even more promising. One of the major disadvantages of solar and wind power is their instability. Therefore, the supply of renewable power into the main grid can sometimes be problematic and some losses of green power are possible due to rejection by the transmission network operators. As a result, there may be cases where it will be more viable financially to convert the

* Corresponding Author's Email: vytautas.adomavicius@ktu.lt.

produced green electricity into hydrogen and supply it to a variety of consumers. The concept of the proposed plant for hydrogen production is explained below by presenting and describing a block diagram of the plant's structural units and a simplified diagram of the control system of technological processes taking place in this plant. All possible and potentially profitable targets for the use of the RES-based power plant and future research directions are also described. The key findings and proposals described in this chapter are summarised in the conclusions.

Keywords: renewable energy, green hydrogen, hydrogen storage and supply into the natural gas pipeline, power balancing, and frequency support in the main power system, artificial intelligence

NOTATIONS AND ABBREVIATIONS

dH_2O	distilled water
H_2	hydrogen
H_2O	water
O_2	oxygen
P_1	power flow from the WT1
P_n	power flow from the WTn
P_D	power flow to the EWD
P_{FC}	power flow from the INV FC
P_{PV}	power flow from the PVPP
P_R	power flow to the rectifier and EL PEM
U~	alternative voltage
U=	direct voltage
DVG	direct vapour generator
DW	distilled water (filling of bottles)
EL PEM	electrolyser with polymer electrolyte membrane
EWD	electric water distiller
FC PEM	fuel cell with polymer electrolyte membrane
FOC	filling of oxygen cylinders
HEV FS	filling station for hydrogen electric vehicles
HPS P2H2	hydrogen production system "Power to H2"

INV FC	inverter for the FC PEM
INV PV	inverter for the PVPP
LCOE	levelized cost of electricity
NGD	natural gas distribution (pipeline)
PTN	power transmission network
PVPP	photovoltaic power plant
PWT	park of wind turbines
R	rectifier
RES	renewable energy sources
S_1	switch for the 1-st wind turbine
S_n	switch for the last wind turbine in the wind park
SAI	system of artificial intelligence
S_C	switch for the compressor
S_D	switch for the electric water distiller EWD
S_{FC}	switch for the inverter of fuel cells INV FC
S_{PV}	switch for the photovoltaic power plant PVPP
S_R	switch for the rectifier R and electrolyser EL PEM
S_S	switch for the connection to the power system
S_T	switch for the transformer T
SWD	solar water distiller
T	transformer
TS	transformer substation
VC	vapour condenser
WD	water distiller
WT_1	1-st wind turbine
WT_n	n-th wind turbine

Introduction

The use of fossil fuels has grown rapidly over the last three centuries in various sectors of the global economy. Such uncontrolled pollution of the environment by polluting fuel has caused large-scale ecological, biological, climatic and humanitarian problems that can no longer be tolerated. As a result, the Green Revolution is gaining momentum in energy, industry, agriculture, transport, building and other sectors of the economy on a global scale.

These changes are not only driven by the desires to have a clean environment, a normal climate and to quit the "smoking" of industrial and transport pollutants that are harmful to health not only for humans. The fact that electricity produced in the RES-based power plants (in solar, wind and hydro power plants) is currently one of the most competitive in the power market is also a very important factor. According to the data of Lazard Levelized Cost of Electricity Analysis 2020 [1] and Company Statista [2], global LCOE of the main power generation technologies in 2020 were as follows:

- Solar power plants – $37 for MWh;
- Wind power plants – $40 for MWh;
- Hydro power plants [2] – $50 for MWh;
- Natural gas power plants – $59 for MWh (combined cycle);
- Coal power plants – $112 for MWh;
- Nuclear power plants – $163 for MWh.

LCOE includes the cost of power plant construction, on-going consumption of fuel (solar, wind, or hydropower plants are free from fuel usage, and for them, these costs are not applicable), operating and maintenance costs and the cost of power plant decommissioning (huge for the nuclear power plants and very small for solar power plants). The decommissioning of nuclear power plants is complicated, dangerous and lasts for years. Solar power plants have the simplest and shortest decommissioning among all types of electric power plants. The indicated LCOE of the PVPP has a clear trend of further decrease. A new world record for the lowest price for solar power has been set on April 28, 2020, when French energy company EDF and Chinese solar company JinkoPower signed an agreement to sell energy from their 1.5 GW solar power plant to Al Dhafra Power Corporation (UAE) for $0.01353 per kWh (or only for \$13.53 per MWh) [3]. Prices for solar and wind power will keep falling in the future, but not so rapidly [4].

Additionally, a huge superiority of solar and other RES-based power plants over fossil fuel power plants can be seen in the paper [5] in the bar chart Figure 3, where the average carbon intensity of power generation is shown depending on the primary source of energy. The carbon intensity is evaluated by the grams of CO_2 equivalent for the unit of generated power (for 1 kWh).

Thus, the advantages of solar and wind power plants in terms of electricity generation are obvious. As a result, the total annual capacity of new solar and wind power plants alone represents up to 80% of the total installed capacity of all new power plants worldwide in recent years. The world economy is shifting towards a green course [6]. Banks in various countries are already refusing to finance projects related to fossil fuels. Recently, a statement about the decision not to finance new projects in the natural gas industry was made by the president of the European Investment Bank, Dr. Werner Hoyer. He declared: “gas is over” [7].

Rural areas are a very good place to install solar and wind power plants. This was confirmed after German scientists from Fraunhofer ISE proposed agrivoltaic systems for the double use of land [8], and after solar power prices fell sharply in the first decade of the 21st century. Studies of the agrivoltaic systems have shown that land use efficiency in these systems increases by up to 186%. Many countries around the world have begun to develop agrivoltaic systems in rural areas, including those with a very small share of arable land [9].

But solar and wind farms don’t just have good features. One of their major drawbacks is the intermittence and instability of the generated power depending on meteorological conditions. These disadvantages can be eliminated by using power storage equipment for supporting the power plants when necessary. Therefore, the integration of larger amounts of unstable power into the electricity system requires a higher capacity of energy storage facilities. Sometimes electricity transmission grid operators can ask to reduce the output of solar power plants or totally disconnect them. In this case, losses of green power are inevitable. The integration of renewable power into the electric grid is becoming increasingly difficult [10, 11] when the amounts of electricity generated by solar and wind power plants increases significantly. To address such problems, the green electricity generated could be converted to hydrogen energy. Green hydrogen is projected to be an important source for future energy and transport [12, 13] in the world, especially the EU countries. The main obstacle on the way to achieving this goal presently is the high price of green hydrogen. Solar and wind energy are very important and competitive sources of energy for the widespread implementation of the idea of affordable green hydrogen production in Europe and worldwide.

The Renewable Hydrogen Coalition was established in Europe in November 2020 by SolarPower Europe, Wind Europe and Breakthrough Energy organizations for close cooperation in this area. Large-scale hydrogen production and use projects have already been launched in the European Union and in some other advanced countries [14-17]. It is expected that in the near future the price of green hydrogen will be significantly reduced – at least to 1.0-1.5 €/kg. The rapidly declining prices of electricity generated by solar and wind power plants give hope that this goal will be achieved fairly soon.

This chapter presents the concept on the RES-based multipurpose plant for hydrogen production. In this system, hydrogen can be produced in all cases when the energy produced in the wind farm and the solar power plant cannot be supplied into the grid of the electricity system, or when it is not profitable enough to supply the power. Hydrogen production in this plant is one of the most important goals. The concept of this system provides a few possibilities of the produced hydrogen realization, which is described in this chapter.

Concept of RES-Based Plant for Hydrogen Production

The structure of the proposed multipurpose hydrogen production plant is presented in Figure 1. The proposed multipurpose plant for hydrogen production consists of the the following main subsystems: photovoltaic power plant PVPP, park of wind turbines PWT, hydrogen production subsystem HPS P2H$_2$, hydrogen-based electric vehicles filling station HEV FS, transformer substation TS, solar water distiller SWD, electric water distiller EWD and compressor for hydrogen supply into the natural gas distribution (NGD) pipeline.

The hydrogen plant subsystems mentioned above are described in more detail in the following subdivisions of this chapter. The aim of these descriptions is to show all of the possibilities and functional flexibility of the plant's equipment to operate efficiently without significant loss of electric power potentially possible to produce by means of the PVPP and PWT used in this hydrogen plant.

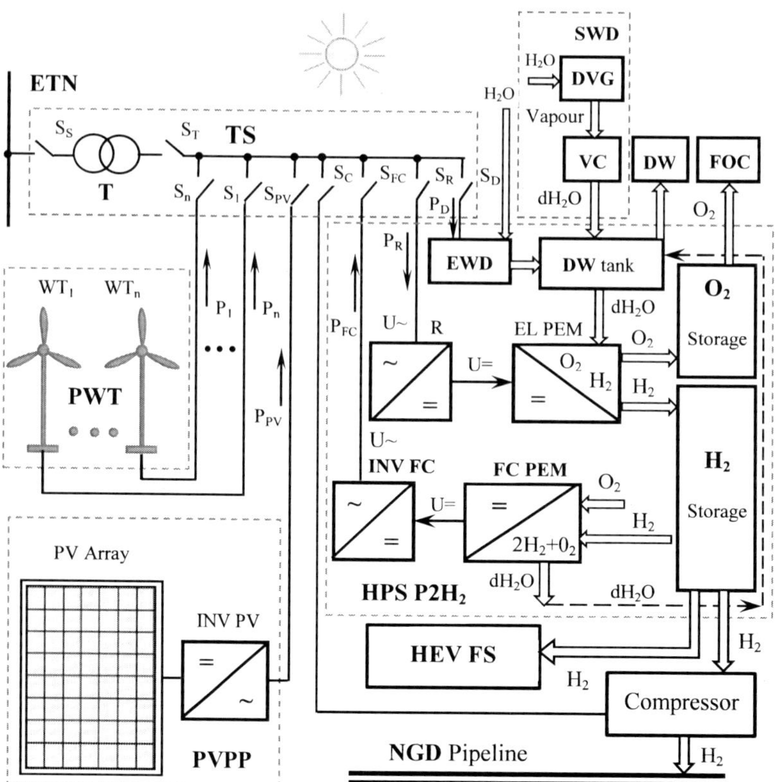

Figure 1. Structure of the multipurpose hydrogen production plant.

Rural areas are a suitable place to build competitive RES-based wind and solar power plants as well as green hydrogen plants. Rural areas tend to have better wind energy resources than suburban areas, and the breakthrough in the construction of solar power plants there is supported by the high potential for efficient dual use of land for electricity and agricultural production. The production of electricity and hydrogen would provide the rural people with a good additional income that would be less affected by natural disasters than the cultivation of various crops.

Power Generation and Utilization in the Plant

Powerful grid-connected PVPP and PWT are operating in close cooperation with the electricity transmission network ETN. As is known, the electricity system network must maintain a balance between the electricity consumption and generation at every moment (the supply of energy into the electricity network must be as much as is required by consumers at that moment). Electricity generation volumes in PVPP and PWT are highly dependent on meteorological conditions. Power balancing processes in the electricity transmission network are regulated by the operator of this network. The regulation scenario of these processes must be prepared in advance. To this end, schedules of power demands by electricity users and schedules of power production by all generators operating in the ETN have to be estimated in advance. The schedules of unstable power generators (such as solar and wind power plants) productivity are made by estimating the forecasts of wind speed and solar irradiance for the next few days (at least for one day ahead). Schedules of power demands and schedules of power production predetermine decisions of the ETN operator. Sometimes, there may be cases when the operator of the network will ask to disconnect one or more rotating wind turbines in the PWT or a certain part of powerful PVPP. Such disconnections can reduce the revenue of RES-based power plant owners and increase environmental pollution when the power plants running on fossil fuel remain operating in the power system.

The RES-based multipurpose power plant for hydrogen production proposed in this chapter never interferes with the generation of electric power and allows utilization of the generated power much more flexibly. This power plant has several possibilities for utilization of the electricity it produces. They can be chosen depending on their profitability. The electricity produced can be used for several purposes, in some cases at the same time. Losses of electric power due to potentially not generated and unused electricity can be avoided at this power plant.

The purposes of the produced electricity consumption that are available in the proposed hydrogen plant are as follows:

- Supply of the electric power from the PVPP, PWT and sometimes from the fuel cells FC PEM into electricity transmission network ETN as a usual electricity producer;
- Participate in the electric power balancing actions of the network operator by supplying the agreed amount of power in the agreed time from the PVPP, PWT, fuel cells FC PEM, and inverter INV FC into the ETN;
- Participate in the frequency support actions of the network operator by supplying the agreed amount of power in the agreed time from the PVPP, PWT, fuel cells FC PEM, and inverter INV FC into the ETN;
- Supply of the electric power from the PVPP and PWT to the rectifier R and electrolyser EL PEM for hydrogen production;
- Supply of the electric power from the PVPP and PWT to the compressor of hydrogen for supplying it into the natural gas distribution pipeline NGD;
- Supply of the electric power from the PVPP and PWT to the back-up distiller EWD used for the production of distilled water.

In this way, PVPP and PWT, which up to now are usually one of the possible causes of power imbalances in the electricity transmission network, can contribute to power balancing and frequency support actions in the ETN in the case where it will be beneficial for the hydrogen plant.

Production of the Distilled Water

The distilled water production subsystem exploits a solar water distiller SWD consisting of a direct vapour generator DVG and a vapour condenser VC. SWD is the main water distiller. The distilled water it produces runs into the tank for distilled water (dH2O tank) for storage. The electric water distiller EWD is used as a backup measure in an emergency when the water level in the dH2O tank is low and there is no sufficient solar radiation. The distilled water production subsystem can operate autonomously, automatically, and independent from the operation of other plant subsystems. The most important tasks for this subsystem are to maintain a sufficiently high water level in the dH_2O tank and to monitor the supply of undistilled water used for the distillation.

Distilled water in the proposed power plant is necessary for hydrogen production. To keep the price of hydrogen as low as possible ($ 1.5 per kg of H_2 or below), distilled water is worth producing on site. Locally produced solar and wind power can be used for this purpose, and possibly it would be most efficient to use free solar thermal energy for the water distillation.

Usual solar water distillers for the production of drinking water are unsuitable for the proposed hydrogen plant due to their low capacity [18-20]. Very high levels of green hydrogen production and usage are forecast for the future. Thus, demand for distilled water is also expected to increase significantly in the near future. Therefore, hydrogen production plants will need to have industrial water distillers with high productivity and a highly efficient operation. Presently, the price of distilled water is below 0.50 USD per kg. It is not a good price for hydrogen production when the target price of green hydrogen is anticipated to be about 1.0-1.5 USD/kg.

Examples of the direct use of solar thermal energy for steam production can be found in scientific publications. Direct solar steam generation is applied in solar thermal power plants. In these power plants, low-parameter steam is usually produced using parabolic concentrators of solar irradiance [21-23]. Direct solar steam generation inside an evacuated tube absorber can also be implemented [24]. High or medium steam parameters are not required for a solar thermal distiller. This distiller should operate more efficiently by receiving low-parameter steam. The scientific publications on industrial solar water distillers are difficult to find. e.g., the title of the publication [25] writes about water distillation for laboratories and industry, however, useful information on high-capacity distillers operating on solar heat is not provided. If currently available industrial solar distillers are not sufficiently adapted to produce water for very large hydrogen plants, they will need to be developed in the near future. The utilization of waste heat in some large energy-intensive plants for the production of distilled water may also be considered for the creation of an efficient modern water distiller.

Hydrogen Production, Usage and Realisation

The green hydrogen in this plant is produced from distilled water by electrolysis using electrolysers with a polymer electrolyte membrane (EL PEM) fed with direct electric current from rectifier R. It is also possible to use several electrolysers connected in parallel. It is also possible to use several electrolysers connected in parallel. The produced hydrogen is supplied into the tank H2 Storage (Figure 1).

Usage and realization of the hydrogen produced in this plant is possible in a few ways:

- Supplying of hydrogen into the natural gas distribution pipeline by means of a compressor;
- Filling of fuel tanks of hydrogen-based electric vehicles, trucks, and agricultural machinery (in the future);
- Production of electric power by the FC PEM, inverter INV FC, and supplying it into the ETN.

In this way, the hydrogen-based power generation may be useful when the proposed plant is involved in power balancing or frequency maintenance activities in order to increase the amount of power supplied to the ETN or to smooth out significant variations in the supplied power. A sufficient number of fuel cells FC PEM and inverters INV FC have to be used in the system for the realization of such functions. Adjustment of the power amount supplied to the ETN is possible by continuously adjusting the number of switched fuel cells FC PEM and inverters INV FC. Usually it is not necessary to pursue all three purposes simultaneously. Only the most beneficial purposes can be chosen.

Oxygen O_2 is a by-product of hydrogen production. It is a useful chemical element used in medicine, aviation, rocket engines, metal cutting, and welding works. Therefore, it is recommended to store the oxygen obtained during the water electrolysis process in the O_2 Storage container, to fill the cylinders of several sizes with oxygen in the post FOC and to sell them on the oxygen market.

Matching of the Capacities of the Plant's Technological Equipment

All hydrogen plant equipment must be matched to each other in terms of power. The most important power generators in the plant are the park of wind turbines PWT and the solar power plant PVPP. Following the choice of the combined capacity of the two types of electricity generators, it is recommended to distribute this capacity by type in such a way that the combined capacity of the two power sources is as constant as possible throughout the year. Initially, it is possible to estimate how much electricity will be generated each month by wind and solar power plants of the same capacity (let us say 1 MW) per year with local perennial average wind and solar energy resources in the selected plant construction site. After that, it is necessary to find the optimal power ratio of PWT and PVPP, at which the fluctuations of the total energy produced during the year, will be the smallest. This can be done by keeping the power and annual energy production of one type of power plant constant (let's say, PWT) and the other type of power and energy production alternating step-by-step until the total variation in energy production of both types of power plants over the year is the smallest. In this case, the calculated power ratio of PWT and PVPP will be that at which the variations in total power over time will approximately be the smallest. On the other hand, the total capacity of PWT and PVPP depends on the projected amount of distilled water to be produced per year and on the amount of electricity to be supplied to the ETN per year. Based on these data, it is possible to calculate the necessary power of electrolysers EL PEM, power of fuel cells FC PEM, volumes of containers H_2 storage, O_2 Storage and dH_2O tank, capacities of solar water distiller SWD, electrical water distiller EWD, and compressor for the produced hydrogen supply into NGD pipeline. For simplicity, only one EL PEM and one FC PEM are shown in Figure 1, but their actual figures and capacities will depend on the capacity of the hydrogen plants to be designed.

Green hydrogen plants can be different in size depending on their capacity. The scale of future RES-bases hydrogen production plants will depend on the hydrogen demand in the region around the site of the plant. It would be most beneficial to build these plants in areas where there would be many hydrogen consumers nearby. It is predicted that in the near future such consumers will be electric planes, ships, trains, trucks, powerful agricultural machinery, which will generate electricity onboard using fuel cells. The supply of hydrogen into the NG pipeline will be allowed temporarily until natural gas is used for various purposes as a carrier of energy. The gas mixture in the NG pipeline will become less harmful to the environment, as no pollutants are produced when hydrogen burns. Very-large-scale green hydrogen plants are already being designed to be built near large ports and to supply hydrogen as fuel. They will have only one purpose – to produce hydrogen as green fuel [26]. Special hydrogen vessels are already being designed for this purpose. One type of vessel will transport the compressed hydrogen ($C\text{-}H_2$) over short distances; other vessels will transport liquefied hydrogen over medium distances ($L\text{-}H_2$) and the rest – over long distances in the form of ammonia (NH_3). Ammonia is used as a hydrogen carrier, a convenient form for delivery. Upon delivery of ammonia to the destination, the technology of cracking and purification of recovered hydrogen will be used in order to make it suitable for use in PEM-type fuel cells for electricity generation. According to the newest study carried out by Fraunhofer institute, only in Germany hydrogen demand is set to grow from 80 TWh in 2030 to 400-800 TWh by 2050 [27].

Control System of the Technological Processes

The control system of the technological processes is a very important part of the multipurpose power plant for hydrogen production. It monitors the status of the operation (on, off) of all equipment used in the plant, monitors and records the instantaneous power generated by all operating wind and solar power plants, the hourly and daily electricity production, the quantities of electricity consumed by the electric water distiller EWD and electrolyser EL PEM, the distilled water production and consumption, the availability of the raw water, the status of operation, the amounts of distilled water produced by the solar water distiller SWD and the electric water distiller EWD, the current amount of distilled water in the dH_2O tank, the amounts of the distilled water released over the DW post for filling of bottles, the production and the consumption of hydrogen and oxygen quantities, their quantities in the containers H_2 Storage and O_2 Storage, the quantities of oxygen released over the post FOC, the quantities of the hydrogen supplied into the NGD pipeline by means of the compressor, the quantities of hydrogen used by the fuel cells FC PEM for electric power production and the quantities of this power supplied into the electricity transmission network ETN over the inverter(s) INV FC, the quantities of hydrogen released over the filling station for hydrogen electric vehicles HEV FS, and other parameters, which have to be evaluated in order to control the technological processes taking place in the plant and assess whether they are going smoothly and efficiently.

The monitoring of technological processes in the plant will require the use of many sensors for various purposes, meters of electric power, water, and gas volumes. Digitized data from all sensors and meters are entered into a computerized artificial intelligence system SAI as shown in the simplified block diagram of the control system in Figure 2.

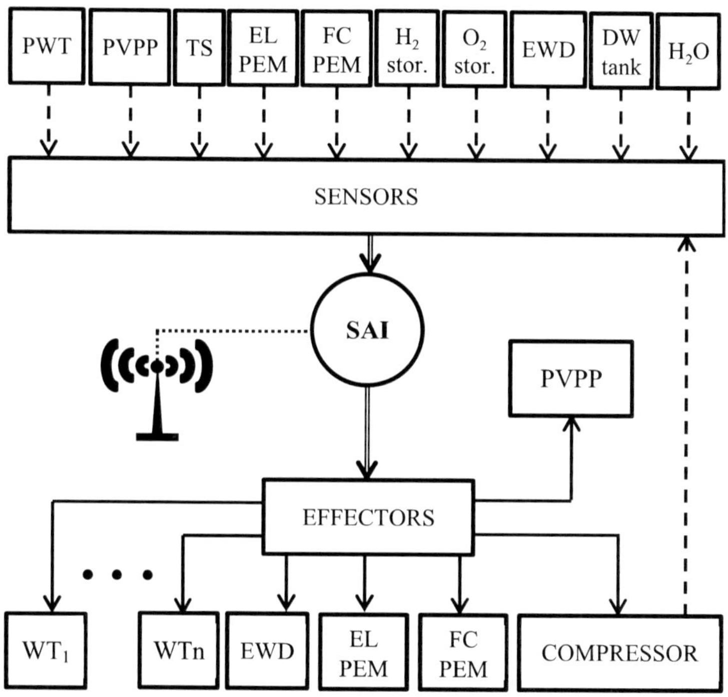

Figure 2. Block diagram of the plant's technological process control system.

A number of other data needed to control the processes in the plant are received over the Wi-Fi channels. Wind speed and solar irradiance data forecasts are required to predict the volumes of electricity generation in the wind and solar power plants and distilled water production in the distiller SWD. All data after entering into the artificial intelligence system SAI are processed according to many algorithms and the factory "brain" submits its proposals. Decisions made by the process control system at the plant have to be supervised and approved by the operator. The operator may not interfere in the work of the SAI if he agrees with the proposed solutions.

The effectors then are turning on or off the necessary technological equipment and most of the processes in the plant are running automatically, except for filling the cylinders of distilled water, oxygen, and hydrogen-powered cars. The reliable and efficient operation of the plant will depend on the quality of the development of the technological process control algorithms.

When building a new green hydrogen plant, some of the objectives described here can be abandoned if there are no such needs in the area. In this case, the structure of the hydrogen plant could be fairly simple. For example, this plant could perform fewer functions: producing electricity and supplying it to the electricity grid when this is economically beneficial, producing and supplying green hydrogen to the main natural gas pipeline, and realizing in the market the byproduct (oxygen) when there is no demand for power in the main electrical power system.

This type of plant also could choose to perform other functions: to produce and store hydrogen and oxygen when there is excess power in the main electricity network, to participate in power balancing and frequency support in the main electricity grid, and to realize the hydrogen and oxygen unused for power balancing and frequency support.

Further Research Directions

The most important directions for further research on the way to the creation of cost-effective RES-based plants for the production of green hydrogen are following:

- Elaboration of the cost-effective solar water distiller based on the solar collectors for water heating;
- Creation of an efficient and reliable system for control of the technological processes taking place in the plant;
- Minimization of the total price of the whole system in order to have a low price of the green hydrogen.

The creation of an efficient and reliable system for control of the technological processes taking place in the plant will require a novel knowledge for the evaluation and processing of many variable predicted parameters as well as in a real-time because of the integration of the RES-based power plants. Creating this control system can benefit from the results of the latest research and scientific information on the renewable power generation, sustainability, intelligent computing, artificial intelligence, optimization, Industry 4.0, and other useful information presented in the references [27-29].

CONCLUSION

In the competitive struggle between electricity generation technologies, the breakdown of solar and wind power plants into the first places according to the Levelized cost of electricity created the conditions to overcome the last obstacle on the way to the creation of a green economy – to develop competitive green hydrogen production technologies. The need for low-cost green hydrogen as an energy carrier is enormous in today's world. Electricity and thermal energy producers, powerful vehicles, and industry will be the largest hydrogen users. One of the possible concepts of a novel multipurpose plant for hydrogen production based on the most promising solar and wind power today is described in this chapter. The instability of solar and wind power is one of the major drawbacks because the supply of this power into the grid of electric power system may not be welcome in cases when the total power of operating generators in the system exceeds the total electric power required by all electricity consumers connected to the system. Considerable losses of green power are possible in this case if the operator of the power transmission network asks to disconnect one or more rotating wind turbines in the PWT or a certain part of powerful PVPP. Such disconnections can reduce the revenue of RES-based power plant owners and increase environmental pollution when the power plants running on fossil fuel remain operating in the power system. Besides, it would be a waste of green electricity production potential. The proposed concept of RES-based multipurpose power plant for hydrogen production presents a few options for utilization of the electric power produced by the plant in a cost-effective way using the eco-friendly technologies. The plant will have the possibility to choose one or more ways to realize the self-produced electricity at the same time. Losses of electric power due to potentially not generated and unused electricity can be avoided at this power plant. The most important features of the novelty of this concept are the possibility of avoiding possible losses in the production of electricity and choosing the most profitable ways for realizing of that "unnecessary" energy.

Acknowledgments

The author is very grateful to *Valeriy Vladimirovich Kharchenko, Professor of Federal Scientific Agroengineering Centre, Russia*, for the invitation and encouragement to prepare this chapter for this book. My sincere thanks also go to all the editors of this book for the comments and advice that have been helpful in improving the quality of the chapter.

References

[1] Carroll, D. *Biden plans to fight climate change in a way no U.S. president has done before*. January 20, 2021. Online [05.02.2021]. Available at: https://www.e-mc2.gr/el/news/biden-plans-fight-climate-change-way-no-us-president-has-done.

[2] *Levelized cost of electricity (LCOE) of hydropower worldwide from 2010 to 2019*. Company Statista. Online [05.02.2021]. Available at: https://www.statista.com/statistics/799349/lcoe-of-hydropower-worldwide/.

[3] Bellini, E. Abu Dhabi's 1.5 GW tender draws world record low solar bid of $0.0135/kWh. *PV Magazine*. April 28, 2020. Online [07.01.2021]. Available at: https://www.pv-magazine.com/2020/04/28/abu-dhabis-2-gw-tender-draws-world-record-solar-bid-of-0-0135-kwh/.

[4] Dorr, A; Seba, T. *RethinkX Sector Disruption.*, (2021), February 2021, 32 p. Available at: https://www.rethinkx.com/energy.

[5] Adomavicius, V; Kaminickas, M. *Energetika. Analysis of PV Power Future Development Possibilities.*, 2014, vol. 60, issue 4, pp. 233-248. Available at: https://www.researchgate.net/publication/271836686_Analysis_of_PV_power_future_development_possibilities.

[6] Arbib, J; Seba, T. *Rethinking Humanity.*, (2020), June 2020, 89 p. Available at: https://www.rethinkx.com/humanity.

[7] Queally, J. Climate Leaders Celebrate as European Investment Bank Chief Declares 'Gas Is Over.' *EcoWatch.*, Online [28.01.2021]. Available at: https://www.ecowatch.com/european-investment-bank-fossil-fuels-2650067861.html?rebelltitem=4#rebelltitem4.

[8] Schindele, S; Trommsdorff, M; Schlaak, A. etc. *Applied Energy.* Implementation of agrophotovoltaics: Techno-economic analysis of the price-performance ratio and its policy implications. 2020, vol. 265, 19 p. Available at: https://doi.org/10.1016/ j.apenergy.2020.114737.

[9] Bellini, E. Israeli government wants to boost development of agrivoltaics. *PV Magazine.*, February 1, 2021. Online [02.02.2021]. Available at: https://www.pv-magazine.com/2021/02/01/israeli-government-wants-to-boost-development-of-agrivoltaics/.

[10] Rakhshani, E; Rouzbehi, K; Sánchez, AJ. etc. *Energies.* Integration of Large Scale PV-Based Generation into Power Systems: A Survey., 2019, vol. 12, 1425, 19 p. Available at: https://www.mdpi.com/1996-1073/12/8/1425.

[11] Impram, S; Nese, SV; Oral, B. *Energy Strategy Reviews.* Challenges of renewable energy penetration on power system flexibility: A survey. 2020, vol. 31, 12 p. Available at: https://doi.org/10.1016/j.esr.2020.100539.

[12] *New Renewable Hydrogen Workstream.* SolarPower Europe. February 26, 2021. Available at: https://www.pveurope.eu/ markets-money/solarpower-europe-new-renewable-hydrogen-workstream.

[13] Bellinni, E. Novel algorithm for integrating solar, wind, hydrogen. *PV Magazine*. February 9, 2021. Online [12.02.2021]. Available at: https://www.pv-magazine.com/2021/02/09/novel-algorithm-for-integrating-solar-wind-hydrogen/.

[14] Enkhardt, S. More green hydrogen for Port of Rotterdam. *PV magazine*. February 9, 2021. Online [14.02.2021]. Available at: https://www.pv-magazine.com/2021/02/09/more-green-hydrogen-for-port-of-rotterdam/.

[15] Molina, PS. New European project to drive 95 GW of solar and 67 GW of hydrogen. *PV Magazine*. February 12, 2021. Online [15.02.2021]. Available at: https://www.pv-magazine.com/2021/02/12/new-european-project-to-drive-95-gw-of-solar-and-67-gw-of-hydrogen/.

[16] Carroll, D. Australia's first hydrogen fuel cell factory. *PV Magazine*. February 22, 2021. Online [25.02.2021]. Available at: https://www.pv-magazine.com/2021/02/22/australias-first-hydrogen-fuel-cell-factory/.

[17] Peacock, B. Australian green hydrogen vessel deemed 'highly competitive'. *PV Magazine*. March 02, 2021. Online [03.03.2021]. Available at: https://www.pv-magazine.com/2021/03/02/australian-green-hydrogen-vessel-deemed-highly-competitive/.

[18] Kathum, J; Hamed, J; Abdul-Munaim, A. *International Journal of Computation and Applied Sciences*. Design and construction of water solar distillation, work by steam circulation technique., 2017, vol. 3 (2), pp. 239-245. Available at: https://www.researchgate.net/publication/320416858_Design_and_construction_of_water_solar_distillation_work_by_steam_circulation_technique.

[19] Mehta, A; Vyas, A; Bodar, N. *International Journal of Advanced Science and Technology*. Design of Solar Distillation System. 2011, vol. 29, pp. 67-74. https://www.researchgate.net/publication/285880564_Design_of_solar_distillation_system.

[20] Bahuguna, S; Verma, P; Chauhan, S. *NCETEST*. A Review and Recent Developments in Solar Water Distillation., 2014, 6 p.

Available at: https://www.academia.edu/11933167/ A_Review_and _Recent_Developments_in_Solar_Water_Distillation.

[21] Nihaj Uddin Shan ANM. Direct steam generation in solar collectors. *Technical Report*. Islamic University of Technology. 2020, 28 p. Available at: http://DOI:10.13140/RG.2.2.19141.47847.

[22] Fernandez-Garcıa, A; Zarza, E; Valenzuela, L. ect. *Renewable and Sustainable Energy Reviews*. Parabolic-trough solar collectors and their applications., 2010, vol. 14, pp. 1695-1721. Available at: http://doi:10.1016/j.rser.2010.03.012.

[23] Sait, HH; Martinez-Val, JM; Abbas, R. etc. *Applied Energy*. Fresnel-based modular solar fields for performance/cost optimization in solar thermal power plants: A comparison with parabolic trough collectors., 2015, vol. 141, pp. 175-189. Available at: http://dx.doi.org/10.1016/j.apenergy.2014.11.074.

[24] Bataineh, KM; AL-Karasneh, AN. *AIMS Energy*. Direct solar steam generation inside evacuated tube absorber., 2016, vol. 4 (6), pp. 921-935. Available at: http://DOI:10.3934/energy.2016.6.921.

[25] Joshi, SB. *Proceedings of the PEPCCI conference*. New techniques for distillation of water for laboratories and industry. Vasad, 2012, 6 p. Available at: https://www.researchgate.net/publication/ 287978113_NEW_TECHNIQUES_FOR_DISTILLATION_OF_WATER_FOR_LABORATORIES_AND_INDUSTRIES.

[26] Peacock, B. Overlooked supply chain enters Australia's green hydrogen export mix. *PV Magazine*. March 17, 2021. Online [18.03.2021]. Available at: https://www.pv-magazine.com/2021/ 03/17/overlooked-supply-chain-enters-australias-green-hydrogen-export-mix/.

[27] Hannen, P. German hydrogen demand likely to pick up from 2030, says Fraunhofer. *PV Magazine*. June 7, 2021. Online [08.07.2021]. Available at: https://www.pv-magazine.com/2021/06/ 07/german-hydrogen-demand-likely-to-pick-up-from-2030-says-fraunhofer/.

[28] *Proceedings of the 1st International Conference on Intelligent Computing and Optimization 2018 (ICO'2018)*. Editors: Vasant P.,

Zelinka I., Weber G.-W. Available at: https://www.springer.com/gp/book/9783030009786.

[29] *Proceedings of the 2nd International Conference on Intelligent Computing and Optimization 2019 (ICO 2019)*. Editors: Vasant P., Zelinka I., Weber G.-W. (Eds.) Available at: https://www.springer.com/gp/book/9783030335847.

[30] *Proceedings of the 3rd International Conference on Intelligent Computing and Optimization 2020 (ICO 2020)*. Editors: Vasant P.; Zelinka I.; Weber G.-W. Available at: https://link.springer.com/book/ 10.1007/978-3-030-68154-8.

In: Advances of Machine Learning … ISBN: 978-1-68507-211-7
Editors: Pandian Vasant et al.

Chapter 2

DEVELOPING A BAYESIAN NETWORK TO MODEL ENVIRONMENTAL, ORGANIZATIONAL, AND HUMAN RISK FACTORS: A CASE STUDY ON WIND TURBINES

Maryam Ashrafi**, PhD
Industrial Engineering and Management Systems Department,
Amirkabir University of Technology, Teheran, Iran

ABSTRACT

Technological systems composed of technical parts embedded in human, organizational, and environmental contexts have various interactions among their components. Almost all traditional risk assessment techniques, such as Failure Modes and Effect Analysis (FMEA) and Fault Tree Analysis (FTA), are subject to some limitations in terms of establishing efficient links between risk models and human and organizational factors for studying modern technological systems.

This chapter proposes an integrated risk assessment framework for technological systems using a Bayesian Network (BN) so that different

* Corresponding Author's Email: ashrafi.mm@aut.ac.ir.

levels of a system and their interactions can be considered. The mentioned framework is based on a systemic approach of incorporating all levels of the system and their interactions. The proposed model was applied to risk modeling in a wind turbine as an instance of complex power generating systems consisting of several structural, electrical, and mechanical components interacting with human resource and organizational and environmental factors.

Different scenarios of risk analyses were investigated and it was found that a BN model was an effective tool for understanding the system's behavior and assessing risk and reliability beyond common technical factors. The results pointed to the role and significance of organizational and human reliability factors in such a system which could be used in decision-making concerning the maintenance strategy and crew working conditions in a cognitive-based approach.

The developed BN model and the method of risk analysis applied in this study may be extended to the entire wind farm or other complex technological systems. It can be utilized as a powerful decision support system to help senior management understand the risks targeting complex systems based on a systemic approach.

INTRODUCTION

Interactions among the components of technological systems increase the risk of failure in such complex technological systems. Complex systems are composed of technical subsystems embedded in human, organizational, and environmental contexts with various interactions and nonlinear relationships among their components. They are open to the environment and make exchanges with it (Woo and Lee 2012). These characteristics result in changing risks in technological systems, and consequently a variety of existing risk assessment approaches are inefficient to cope with risks (Adarsh and Janga Reddy 2013).

Continuous population growth and effort for improving the living standards have led to increasing energy demand all over the world. So, there is an essential need for increasing the use of renewable energy sources in the energy portfolio of each country. Wind turbines are a renewable source of energy expected to play an increasingly important role in providing electricity in the near future (Guo, et al., 2009). Wind turbines act as instances of complex systems in which technical, human, organizational, and environmental factors interact with each other. A survey of the history of accidents in this field shows that most major power grid blackouts that occurred in the past were initiated by a single event (or multiple related events), leading to a catastrophic accident or even entire system collapse. It is very critical to assess and control risks of wind turbines for three reasons: 1) wind turbines have an important role in providing electricity due to the increase in capacities and the number of grid-connected wind turbines (Guo, et al., 2009), 2) wind power penetration in power systems increases at a significant rate and it is estimated to be the fastest growing renewable energy resource (Global Wind Energy Council (GWEC) 2012), and 3) the profitability of wind farms is affected by system availability and reliability (O&M costs).

Traditional risk assessment approaches consider a system as a composition of independent elements that can be easily understood by breaking them down to their smallest elements and studying the manner of their elements (Ackoff 1987). Almost all traditional risk assessment methods such as Failure Modes and Effect Analysis (FMEA), Hazard and Operability Analysis (HAZOP), Fault Tree Analysis (FTA), and Probabilistic Risk Analysis (PRA) focus on technical factors and rely on a chain of linear cause and effect analyses (Ramesh and Saravannan 2011, Alzbutas, et al., 2014). These techniques are not able to model an efficient link between technical factors and human, organizational, and environmental factors (Mirzaei Aliabadi, et al., 2019). Therefore, the risk analysis of wind turbines as complex power generating systems consisting of several structural, electrical, and mechanical components interacting with human resource and organizational factors within natural, political, and economic environments is a significant factor in ensuring the success of a wind power system (Janghorbani, et al., 2014, Gheydi, Effatnejad and Ramezanpour 2014, Ashrafi, Davoudpour and Khodakarami 2015).

Today, to cover traditional methods, several innovative computing and novel optimization techniques have been developed with applications in many research areas such as energy and risk modeling (Vasant, Zelinka and Weber, Intelligent Computing & Optimization 2019, Vasant, Zelinka and Weber, Intelligent Computing and Optimization 2020, Vasant, Zelinka and Weber, Intelligent Computing and Optimization 2021). In this chapter, the factors affecting reliability and safety of a wind turbine are presented and the impact of organizational, human, and environmental factors on the emergent system risk level is investigated. To achieve the mentioned objectives, a Bayesian Network (BN) model is developed for the risk analysis of a wind turbine.

The developed model is capable of modeling interactions among the elements of complex systems in four layers of technical, environmental, organizational, and human factors and their interactions. This chapter proposed an integrated risk assessment framework for complex technological systems and developed a BN to demonstrate its potential applicability for modeling risk and reliability in complex systems. Furthermore, the model analyzed the reliability of each component of the considered system considering the impacts of organization safety culture, maintenance strategy, and human error probability as well as climate states.

This chapter is organized as follows: Section 2 briefly describes a wind turbine structure. Section 3 introduces the background, concept, and application of BN models. Section 4 elaborates the development of a BN for risk analysis of wind turbines. Section 5 presents the model implementation for wind turbines and discusses the results. Section 6 presents the conclusions.

THE STRUCTURE OF WIND TURBINES

A wind turbine includes eight major components which are rotor, nacelle, tower, gearbox, generator, pitch system, yaw system, and the foundation of the system. The kinetic energy extracted from the wind is converted to mechanical energy by the rotor and rotating blades. Generator converts mechanical energy to electricity (Ashrafi, Davoudpour and Khodakarami 2016).

Nacelle is a subsystem at the top of the tower which is the main structure and contains the main components such as turbine generator, gearbox, drive train, electronic control system, and grid interface (Ribrant 2006, Saeed 2008). The drive train is composed of different mechanical elements such as bearings and shaft. Its main function is to interconnect the hub, the gearbox, and the generator (Ribrant 2006, Sterzinge 2004).

A wind turbine gearbox transforms low-speed revolution into high-speed revolution. It is connected to an electric generator through a high-speed shaft (Ribrant 2006), transforming the rotational mechanical energy to electrical energy. The generator output is then connected to the electrical grid station for electrical energy supply.

The main function of pitch and yaw systems is to adjust blades and stop turbine performance during inappropriate climate and storm. The whole structure is supported by a tower fastened to wind turbine foundation and raises nacelle to a height where maximum wind speeds can be extracted (Sterzinge 2004, Saeed 2008).

BN (Background, Concepts and Applications)

Because a complex technological system combines different structural and social dimensions, it is essential that methods that can cover both dimensions be integrated. BN model is able to integrate the mechanical and social dimensions of systems.

The academic literature in the field of risk analysis and modeling indicates the expansion of the use of BNs. BN is a network-based tool that allows the analyst to display and use different information, both definite and probable, and to display different and complex relationships between a large number of variables. The main origin of this method is artificial intelligence; however, in recent years, in addition to its applications in academic fields, its usage has also expanded in industry. These networks are often used to represent causal relationships in complex phenomena, systems, or processes and to enable the use of expert knowledge alongside historical data.

BN Background

Although Bayesian approach dates back to 1763 and the publication of Thomas Bayes' article, the term "Bayesian" became popular in the 1950s, as mathematicians and philosophers were largely concerned with the concept of probability during the nineteenth century (M. Ashrafi 2021, Fienberg 2006, Khodakarami 2009). During this period, the idea of "inverse probability" (inverse perception of data to parameter) spread in the field of statistical application (Fienberg 2006).

In the early twentieth century, the "classical or frequentist" approach was introduced, in which probability was interpreted on the basis of frequency, and with the advent of the hypothesis test method and confidence intervals following this approach, statistical science faced a comprehensive revolution in theory and practice. This approach, also known as the "classical approach", quickly penetrated various fields of application and replaced the inverse probability in the first half of the twentieth century (Efron 2005).

In the 1950s, modernization of the principles and theory of statistical decision-making led to the development of "subjective probability" and a greater emphasis on Bayesian theorem (Fienberg 2006). In the following decades, the use of Bayesian statistics grew (Mohaghegh-Ahmadabadi 2007), and the modern approach to using this tool intensified in use in the late 1980s following the introduction of Markov Monte Carlo chain methods that made Bayesian calculations possible in real-sized problems (Khodakarami 2009).

BN Concept

Bayes' theorem is used in cases where the state of an entity affects the states of other variables (Khodakarami 2009). Now, if in a situation, several states and variables are related to each other in a chain and graph, the possible inference with respect to the states of variables that are not directly related is the subject of the BN, because conditional probabilities in these situations cannot be calculated by simply applying the Bayesian theorem. BNs, also known as Bayesian belief networks, causal networks, and probabilistic networks, provide a possible inference about a network of variables at an acceptable time (Fienberg 2006, Khodakarami 2009). Also, the graphical nature of these networks provides a better and more complete intuitive understanding of the relationships between variables (Neapolitan 2004).

BN is a Directed Acyclic Graph (DAG) composed of a set of nodes (variables) connected through a set of directed arcs (links). A BN is an instance of causal networks that graphically encodes a multivariate statistical distribution function of a set of conditional independence statements into a compact concise formalism representing a joint probability distribution (Pearl 1988). BN is used for representing uncertain knowledge in probabilistic systems and has been applied to a vast variety of real-world problems.

BN consists of the conditional probabilities between the parent and child variables captured through the directed links using rules of d-separation (Pearl 1988). Conditional Probability Table (CPT) of each child variable is defined in order to link condition states of the child to the parent variables. The set of conditional probabilities allows calculating the joint probability distribution as follows (Pearl 1988):

$$P(x_1, x_2, \dots, x_n) = \prod_{i=1}^{n} P(x_i | Parent\ x_i) \quad (1)$$

In other words, a BN consists of a set of nodes (variables) and a set of directional edges (representing causal effects) between variables (Jensen 1996). Each variable has a definite and finite number of states, and the set of variables and the relationships between them together form a non-circular directional graph.

We argue that BN is a suitable approach to both qualitative and quantitative modelling of risk and reliability in complex systems. Furthermore, BN provides not only a forward (or predictive) analysis but also a backward (diagnostic) analysis, in which the posterior probability of variables can be estimated. Although in most cases variables are defined discretely, continuous variables have been used in some studies. This paper is not limited to the use of discrete variables representing the nodes of the BN.

BN Application

BN is a general, flexible and powerful approach to modeling risk and uncertainty which justifies its increased applications in various fields in recent years (Fenton and Neil 2007). The range of applications of BNs is very wide containing medical diagnosis in the field of health care, investment (Swain and Guttmann 1983), software and hardware fault diagnosis (Fenton and Neil 2007, Heckerman 1996), and risk modelling in energy and transportation systems (Mohaghegh-Ahmadabadi 2007).

There are several reasons for choosing BNs as a modeling tool. In general, a review of the literature points to the ability of this tool to develop risk analysis frameworks in different areas and with the entry of various factors. BNs have minimum requirements for risk modeling, including simplicity and overcoming the problem of data loss or scarcity (Chapman and Ward 2000). In general, the benefits of BNs can be summarized as follows:

- Possibility of causal modeling of variables (Fenton and Neil 2011).
- Ability to integrate with common risk assessment models in the field of technical risk assessment such as fault analysis tree, event tree (Ashrafi, Davoudpour and Khodakarami 2017, Ashrafi and Davoudpour 2019).
- Ability to reason from cause to effect and vice versa, in other words, performing forward analysis (prediction) and backward analysis (diagnosis) (Fenton and Neil 2011).
- Possibility of reviewing and updating models and calculations based on new evidence (Fenton and Neil 2011).

- Making predictions in case of incomplete data set (Fenton and Neil 2011) and the ability to use the opinions of experts in case of lack of data, especially at the organizational level.
- The possibility of combining different types of evidence including subjective beliefs and objective data (Fenton and Neil 2011).
- Transparency of the analysis and decision-making process in contrast to Black box modeling techniques.
- Ability to relate to social and psychological models and theories: Since most of the related social and psychological theories are presented in the form of regression-based methods and mathematically, it is possible to communicate between these regression-based methods with BNs.
- Ability to evaluate the effects of potential changes in variables and their relationships.
- Ability to be applied to real issues and large-scale problems (Agena 2004).
- Ability to visualize information (Agena 2004, Fenton and Neil 2011).

Building a BN

Creating a BN in general and in each domain of application includes the following three steps (Fienberg 2006):

1) Identifying the variables and possible states (situations) of each domain;
2) Identifying the relationships between variables and expressing it in the form of a graphic structure; and
3) Assessing the required probabilities.

These steps in forming a BN network must be taken one after the other and in an iterative process to finally obtain an acceptable network. Also, building a BN always requires a balance among the accuracy, size, and complexity of the model on the one hand and the effort and cost required to build, update, and infer on the other hand.

The graphical nature of BNs makes it easy to delete, modify, and add variables in the model.

After identifying and defining the variables, the next step is to identify the dependencies and relationships between the variables and graphically represent the model. The last step in building a BN is to evaluate the probabilistic values and assign them to the Node Probability Table (NPT). This table actually shows the intensity and strength of the causal dependence between the interconnected nodes, and depending on the type of node (discrete or continuous), it may be a table of possible discrete or continuous distribution.

Common sources of information for extracting the numerical probabilities required are historical data, subject of literature, and related experts. In situations where data is not available, there are several methods for extracting probabilities using the knowledge of experts (Khodakarami 2009).

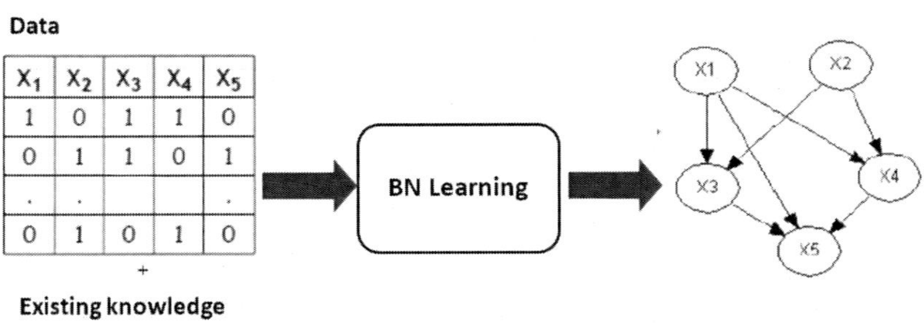

Figure 1. A BN learning process.

After identifying the variables, in order to determine the relationships between the variables and complete the NPTs, it is necessary to determine the parameters and understand the network structure, which is called network learning (parameters and structure). If $\boldsymbol{X} = \{X_1, X_2, \dots, X_n\}$is a set of random variables identified for a model, $\boldsymbol{\mathcal{D}} = \{x^{(1)}, x^{(2)}, \dots, x^{(n)}\}$is the existing data set in relation to these variables and $\boldsymbol{K}$ is knowledge in the field, the purpose of learning the parameters and structure of the network is to determine the values of the parameters and structure (relationships of variables) in a BN so that the consequent BN has the best agreement with D (Fan, Malone and Yuan 2014) (Figure 1).

Developing a BN for Risk Assessment of Wind Turbines

Many technological systems such as wind turbines are composed of technical parts and human and organizational contexts and are interrelated with their environment which lead to an evolving, dynamic, and adaptive behavior. Considering the characteristics of complex technological systems, these systems are analyzed through 4 levels as follows:

1) Technical level including machinery, equipment, and technical facilities. In the proposed model, the overall architecture of the technical subsystem is described using a BN accompanied by the dynamics of the system through the interaction of the variables and agents within and among the system's different levels of environmental, organizational, human, and technical factors;
2) System's human resource with a critical role in risk assessment: Typically, 60%-80% of all accidents are attributed to human error independent of the technological domain of application (Cacciabue 2000, Luxhøj 2003). In the developed model, to study the impact of human factors more realistically, a simulation-based module is used in BN to integrate a cognition-based HRA method (Cognitive Reliability and Error Analysis Method: CREAM) with dynamic models of the technical system behavior and environmental and organization risk factors. This technique uses a set of Common Performance Conditions ((1) adequacy of organization, (2) working conditions, (3) adequacy of MMI and operational support, (4) availability of procedures/plans, (5) number of simultaneous goals or task complexity, (6) available time or adequacy of time, (7) time of day or circadian rhythm, (8)

adequacy of training and experience, and (9) crew collaboration quality) to describe the context in the analyzed scenario/task (Hollnagel 1998).

3) Organizational factors: having a consequential nature and aggregating the effects of human and technical factors following the occurrence of disruption (Mansouri, Nilchiani and Mostashari 2009, Ashrafi and Anzabi Zadeh 2017, Ashrafi, Davoudpour and Khodakarami 2016). In the proposed model, considering wind farm conditions and inspired by models such as ORIM (Qien 2001) and BORA (Aven, Sklet and Vinnem 2006), organizational risk assessment factors of wind turbines are defined. They are defined in 2 categories: (A) Factors related to system structure and culture consisting of safety culture, adequacy of organization, and availability of plans and procedures; (B) Factors related to system maintenance consisting of organization maintenance strategy and quality of performing the required maintenance operations.
4) System environment, including physical environment, regulations, the government, and related unions and associations. The physical environment of a wind turbine and its impact on system is studied in this article.

The presented model assumes complex technological systems like energy production systems as multi-dimensional systems in which environmental, organizational, human, and technical dimensions interact and influence each other. The behavior of components at all the identified levels of a complex technological system can affect (positively, negatively, or neutrally) other components at the same level, other levels, or even the overall system's safety. Furthermore, it is assumed that there is no feedback loop in modeling the interaction of the system's components and all interactions can be modeled with directed arcs in a graph.

The described framework is employed for risk assessment of wind turbines using a BN. A BN is a graphical model that consists of a Directed Acyclic Graph (DAG) with nodes representing variables and directed links between them representing causal relationships. A BN provides a probabilistic framework to quantify causal relationships and interactions under uncertainty. Use of BN makes it possible to examine accident paths in a retrospective backward approach and predict and analyze several likely scenarios in a predictive forward approach. Moreover, given some pieces of evidence or observation, the network probabilities are recalculated to indicate their impact on the whole network.

In this study, the proposed levels of a system and their interactions are modeled and implemented in the commercial BN software package, AgenaRisk.[1] The model represents a wind turbine system along with the factors affecting its risk and reliability at four levels of environmental, organizational, human, and technical factors (Figure 2). The relationships among nodes in all layers of the model are represented by arcs.

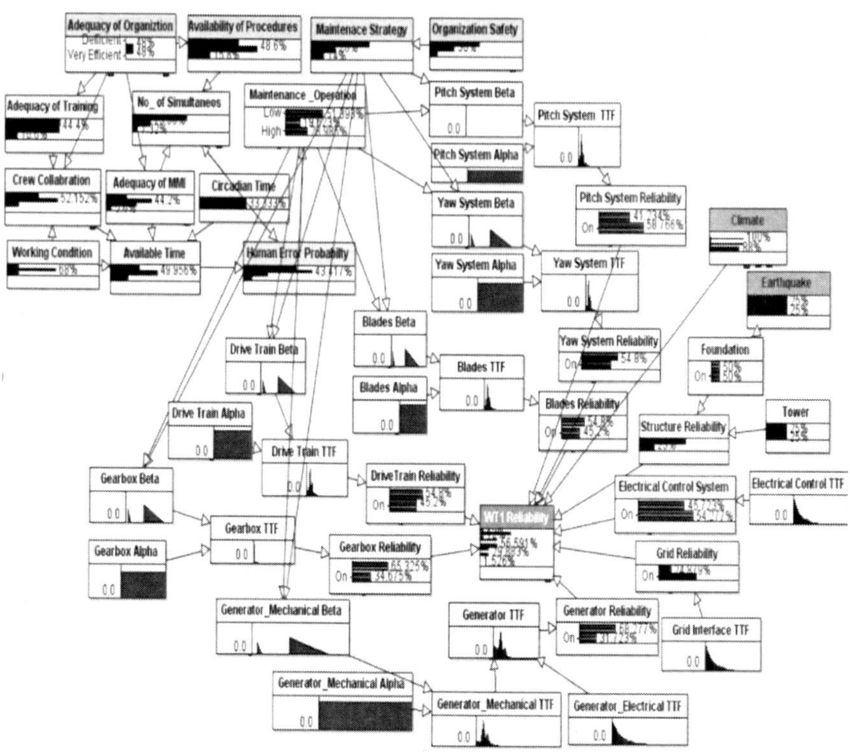

Figure 2. A BN model for a wind turbine consisting environmental, organizational, human and technical risk factors.

[1] Agena Ltd, "AgenaRisk Software", www.agenarisk.com, 2008.

As explained earlier, a wind turbine consists of six major components including rotor, nacelle, tower, gearbox, generator, and foundation of the system which are modeled in the technical layer of BN model. In environmental, organizational, human, and technical segments of the model, a combination of continuous and discrete variables is used. The continuous nodes are mostly time to failure variables of the system components, and discrete variables are used in Label, Rank, or Boolean modes to describe the system behavior (Table 1).

Table 1. Nodes in the BN model of a wind turbine

Node	Type	Modes
Technical Layer		
Time to Failure for Gearbox, Blades, Drive Train, Pitch and Yaw systems	Continuous	Weibull Distribution
Weibull Shape Parameter for Gearbox, Blades, Drive Train, Pitch and Yaw systems	Continuous	Equation with Initial Value
Weibull Scale Parameter for Gearbox, Blades, Drive Train, Pitch and Yaw systems	Continuous	Conditional Experiment (Regarding to Parent Nodes)
Time to Failure for Generator, Grid interface, Electrical Control System	Continuous	Exponential Distribution
Reliability of Generator, Grid interface, Electrical Control, Gearbox, Blades, Drive Train, Pitch and Yaw Systems	Discrete	Binary
Tower	Discrete	Labeled (Crack, Clearance, Resonance, Normal)
Foundation	Discrete	Binary
Structure Reliability	Discrete	Binary
Human Factors Layer		
Working Condition	Discrete	Labeled (Incompatible, Compatible, Advantageous)
Circadian Time	Discrete	Labeled (Night (0:00-8:00), Night (16:00-24:00), Day (8:00-16:00))
Available Time	Discrete	Labeled (Continuously Inadequate, Temporary Inadequate, Adequate)
Adequacy of Man-Machine Interaction	Discrete	Labeled (Inappropriate, Tolerable Adequate, Supportive)

Node	Type	Modes
Number of Simultaneous Goals/ Complexity	Discrete	Labeled (More than Actual Capacity, Matching Current Capacity, Fewer than Actual Capacity)
Crew Collaboration Quality	Discrete	Labeled (Deficient, Inefficient, Efficient, Very Efficient)
Adequacy of Training and Experience	Discrete	Labeled (Inadequate, Adequate with Limited Experience, Adequate with High Experience)
Maintenance Operation Quality	Discrete	Ranked (Low, Medium, High)
Organizational Factors Layer		
Adequacy of Organization	Discrete	Labeled (Deficient, Inefficient, Efficient, Very Efficient)
Safety Culture	Discrete	Ranked (Promotional, Organized, Fully Implemented)
Availability of Procedures and Plans	Discrete	Labeled (Inappropriate, Acceptable, Appropriate)
Maintenance Strategy	Discrete	Labeled (Corrective, Periodic Monitoring, Condition Monitoring)
Environmental Factors Layer		
Climate	Discrete	Labeled (Ice, Extreme Wind Condition, Lightening, Normal)
Earthquake Occurrence	Discrete	Labeled (High, Medium, Low, No Occurrence)

In the technical layer of the BN model, continuous nodes represent time to failure of the basic components and discrete nodes describe the state of the system or subsystems at a particular time. The nodes representing the time to failure for basic mechanical components are completed by Weibull probability density function with different scales and shape parameters. The initial values of Weibull parameters are estimated by considering component mean time to failure and its related standard deviation (Chen, Ho and Mao 2007). These parameters are affected by organization maintenance strategy and maintenance operation quality. Similarly, the time to failure for the basic electrical components of wind turbine such as generator is estimated using exponential probability density function, the parameter of which is affected by organization maintenance strategy and maintenance operation quality.

The BN model allows analyzing the impacts of organizational culture, maintenance strategy, human working conditions, and climate state on the reliability of wind turbine. The existing historical data and expert judgments of the system are used to feed the model. Then, the model is applied to analyze the risk and reliability associated with a wind turbine with regard to the impacts of environmental, organizational and human factors.

Model Implementation and Discussion

This section presents the results of applying a BN to analyze the interactions of different levels of a wind turbine as a system and their impacts on the risk and reliability of the system and its components.

Impact of Environmental Factors on Reliability

The equipment and workers of each wind turbine are directly exposed to the natural environment and may face environmental risks. Environmental risks are caused by dynamic conditions like weather, temperature, and earthquake considered in the developed model. As shown in the model, the reliability of the whole wind turbine is affected by site climate as well as the reliabilities of components in a serial-parallel configuration. In the proposed model, climate is considered as a node, the state of which directly affects the risk levels for the wind turbine. Figure 3 illustrates the underlying risk of wind turbine in different climate states.

As shown in Figure 3, environmental factors such as climate may fundamentally affect the system risk. Hence, models cannot adequately address systems' risk and behavior in case they are restricted to technical factors.

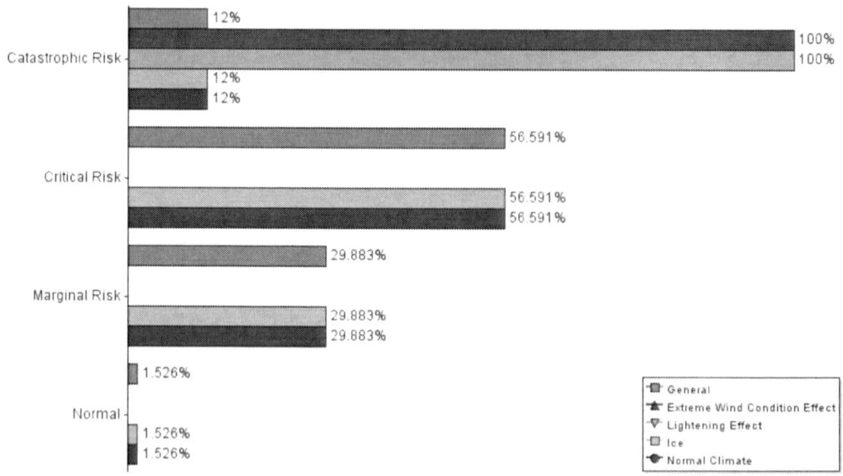

Figure 3. Categorization of risk levels for a wind turbine considering different climate states.

IMPACTS OF ORGANIZATIONAL AND HUMAN FACTORS ON RELIABILITY

To shed light on the undeniable impacts of organizational and human reliability factors on the reliability of system and components, two scenarios are created assuming the prior probabilities of organizational and human factors in their best and worst modes. To perform this analysis, the factors “adequacy of organization”, “organization safety culture”, “working condition”, and “circadian time” are assumed to be in their best and worst modes. The first scenario is analyzed given the “adequacy of organization” in a very efficient mode, “organization safety culture” in a fully implemented mode, “working condition” in an advantageous mode, and “circadian time” on day time of work (8:00-16:00). To perform the second scenario analysis, the factors of “adequacy of organization”, “organization safety culture”, “working condition”, and “circadian time” are assumed in deficient, promotional, incompatible, and midnight working time modes, respectively. These factors are selected from organizational and human factor layers, because they are independent of other factors; also, almost a long period is required to modify or change the states of these factors. On the other hand, other factors associated with the model are significantly affected by the aforementioned factors. Then, the conditional probabilities are recalculated given the pieces of evidence. The result of the described scenarios in terms of the impacts of the studied factors on gearbox reliability is presented in Figure 4. It shows changes in probability distribution and reliability of gearbox's time to failure from Scenarios 1 to 2. The results are described with a focus on gearbox because it is one of the most critical components of wind turbines, which has a significant influence on the time to failure of wind turbine and consequently its reliability.

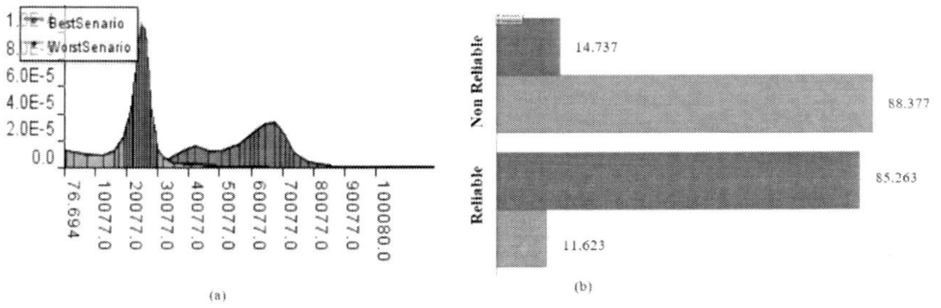

Figure 4. (a) Gearbox time to failure and (b) gearbox reliability for the best and worst scenarios.

As shown in Figure 4, the gearbox's time to failure following Weibull distribution is condensed in Scenario 1, indicating that the scale parameter of Weibull distribution function is significantly affected by the studied factors. The second scenario results in a significant increase in the probability of failure and a considerable decrease in the reliability of the gearbox at some time slices of its lifetime.

The BN provides a possibility for verifying the variation in the gearbox reliability due to the improvement of the related organizational and human reliability factors and performing cost-benefit analyses to ensure the tradeoff among the obtained target and its related cost and time.

CONCLUSION

Many technological systems which are composed of technical parts embedded in human, organizational, and environmental contexts can be categorized as complex systems. In spite of the aforementioned characteristics of complex technological systems, most studies on technological systems relying on traditional risk assessment methods are incapable of modeling interactions among the elements of these systems. They focus on technical risk and reliability analysis and do not address the environmental, organizational, and human factors and their interactions.

This chapter proposed an integrated risk assessment framework for complex technological systems and developed a BN to show its potential applicability for modeling risk and reliability in complex systems considering various system levels and their interaction through a cause-and-effect approach. The author analyzed the reliability of each component of the considered system and the overall system and studied the impacts of organization safety culture, maintenance strategy, and human error probability on reliability as well as climate states.

The result confirmed the significant impact of the main environmental, organizational, and human reliability factors on reliabilities of components. It highlighted the role and significance of organizational and human reliability factors in such a system which could be used in decision-making concerning the maintenance strategy and crew working conditions in a cognitive-based approach. The BN analysis showed that the gearbox reliability was mostly affected by maintenance strategy and organization safety culture. Thus, the maintenance plan should pay special attention to this component. The main limitation of this research which makes research validation challenging is the lack of historical data for some variables in the developed network. If the required data is available objectively, we can revise the model and add conditional probabilities based on objective historical data.

In future works, the reliability analysis applied in this study can be extended to the entire wind turbine or wind farm. In addition, future works may focus on providing an overview of the implementation of a redundancy in the system, which is an example of the BN as a tool to study the feasibility of an action before deployment.

REFERENCES

Ackoff, Russel L. 1987. "Mission statements." *Strategy & Leadership* 15 (4): 30 - 31.

Adarsh, S., and M. Janga Reddy. 2013. "Reliability analysis of composite channels using first order approximation and Monte Carlo simulations." *Stochastic Environmental Research and Risk Assessment* 27:477–487.

Agena . 2004. *Intelligent solutions for quantifying Operational Risk.* Agena White Paper.

Alzbutas, Robertas, Tomas Iesmantas, Mantas Povilaitis, and Jurate Vitkute. 2014. "Risk and uncertainty analysis of gas pipeline failure and gas combustion consequence." *Stochastic Environmental Research and Risk Assessment* 28:1431–1446.

Arabian-Hoseynabadi, H., P. J. Tavner, and H. Oraee. 2010. "Reliability comparison of direct-drive and geared drive wind turbine concepts." *Wind Energy* 13: 62–73.

Ashrafi, M., H. Davoudpour, and V. Khodakarami. 2015. "Risk assessment of wind turbines: Transition from pure mechanistic paradigm to modern complexity paradigm." *Renewable and Sustainable Energy Reviews* 51, 347-355.

Ashrafi, M., and H. Davoudpour. 2019. "A Hierarchical Bayesian Network to Compare Maintenance Strategies based on Cost and Reliability: A Case of Onshore Wind Turbines." *International Journal of Industrial Engineering* 26(3)-p361-375.

Ashrafi, M., and S. Anzabi Zadeh. 2017. "Lifecycle risk assessment of a technological system using dynamic Bayesian networks." *Quality and Reliability Engineering International* 33 (8), 2497-2520.

Ashrafi, M., H. Davoudpour, and V. Khodakarami. 2016. "A Bayesian network based framework to evaluate reliability in wind turbines." *Wind and Structures, An International Journal* 22 (5), 543-553 .

Ashrafi, M., H. Davoudpour, and V. Khodakarami. 2017. "A Bayesian Network to Ease Knowledge Acquisition of Causal Dependence in CREAM: Application of Recursive Noisy-OR Gates." *Quality and Reliability Engineering International* 33 (3)-Pages 479-491.

Ashrafi, Maryam. 2021. "Forward and backward risk assessment throughout a system life cycle using dynamic Bayesian networks: A case in a petroleum refinery." *Quality and Reliability Engineering International* Volume37, Issue1, 309-334.

Aven, Terje, Snorre Sklet, and Jan Erik Vinnem. 2006. "Barrier and operational risk analysis of hydrocarbon releases (BORA-Release)-Part I. Method description." *Journal of Hazardous Materials* 681–691.

Cacciabue, P.C. 2000. "Human factors impact on risk analysis of complex systems." *Journal of Hazardous Materials* 101–116.

Chapman, C., and S. Ward. 2000. "Estimation and evaluation of uncertainty: a minimalist first pass approach." *International Journal of Project Management* v. 18, p. 369-383,.

Chen, S. K., Tin Kin Ho, and B. H. Mao. 2007. "Component reliability estimations without field data." *Hong Kong Institution of Engineers, Transactions* 10-17.

Cheng, Jin, and Q. S. Li. 2009. "reliability based analysis of torsional divergence of long span suspension bridges." *Wind and Structures, An International Journal* 12 (2).

Efron, B. 2005. "Bayesians, Frequentists, and Scientists." *Journal of the American Statistical Association* v. 100, p. 1-5.

Fan, X., B. Malone, and C. Yuan. 2014. "Finding Optimal Bayesian Networks Using Constraints Learned from Data." *Proceedings of the 30th Annual Conference on Uncertainty in Artificial Intelligence (UAI-14)*. Quebec.

Fenton, N. E., and N. Neil. 2007. "Using Ranked Nodes to Model Qualitative Judgements in Bayesian Networks." *IEEE Transactions on Knowledge and Data Engineering* v. 19, n. 10, p. 1420-1432.

Fenton, N., and N. Neil. 2011. *The use of Bayes and causal modelling in decision.* Agena Ltd.

Fienberg, S. E. 2006. "When Did Bayesian Inference Become "Bayesian"?" *Bayesian Analysis* v. 1, n. 1, p. 1-40.

Gheydi, Milad, Reza Effatnejad, and Parviz Ramezanpour. 2014. "Evaluation of Uncertainty in Hybrid Plants, Including Wind Turbine, Photovoltaic, Fuel Cell, and Battery System using Fuzzy Logic." *Indian Journal of Science and Technology* 7 (2: 113–122).

Global Wind Energy Council (GWEC). 2012. *Global Wind Energy Outlook 2012.* Global Wind Energy Council (GWEC).

Guo, Haitao, Simon Watson, Peter Tavner, and Jiangping Xiang. 2009. "Reliability analysis for wind turbines with incomplete failure data collected from after the date of initial installation." *Reliability Engineering and System Safety* 1057–1063.

Heckerman, D. 1996. *A Tutorial on Learning with Bayesian Networks.* Microsoft Research.

Hollnagel, E. 1998. *Cognitive Reliability and Error Analysis Method (CREAM).* First Edition. Oxford: Elsevier Science.

Janghorbani, Mohammadreza, Seyed Mohammad Shariat, Vahid Amir, Mohammad Ghanbari Jolfaei, and Ali Madanimohammadi. 2014. "Risk Management Strategies for a Wind Power Producer in Electricity Markets." *Indian Journal of Science & Technology* 7 (8): 1107–1113.

Jensen, F. V. 1996. *Introduction to Bayesian Networks.* Springer .

Khodakarami, V. 2009. *Applying Bayesian Networks to model Uncertainty in Project Scheduling*, PhD Dissertation . Queen Mary, University of London.

Luxhøj, J. T. 2003 . "Probabilistic Causal Analysis for System Safety Risk Assessments in Commercial Air Transport." *Proceedings of the Workshop on Investigating and Reporting of Incidents and Accidents (IRIA).* Williamsburg, VA. 17-38.

Mansouri, M., R. Nilchiani, and A. Mostashari. 2009. "A Risk Management-based Decision Analysis Framework for resilience in

Maritime Infrastructure and Transportation Systems." *Systems Conference.* 3rd Annual IEEE. 35-41.

Mirzaei Aliabadi, M., R. Esmaeili, I. Mohammadfam, and M. Ashrafi . 2019. "Human Reliability Analysis (HRA) Using Standardized Plant Analysis Risk-Human (SPAR-H) and Bayesian Network (BN) for Pipeline Inspection Gauges (PIG) Operation: A Case Study in a Gas Transmission Plant." *Health Scope* 8(3):e87148.

Mohaghegh-Ahmadabadi, Z. 2007. *On the Theoritical Foundations and Principals of Organizational Safety Risk Analysis (PhD Thesis).* University of Maryland, College Park.

Neapolitan, R. E. 2004. *Learning Bayesian Networks.* Prentice Hall.

Pearl, Judea. 1988. *Probabilistic Reasoning in Intelligent Systems: Networks of Plausible Inference.* Morgan Kaufmann Publishers.

Qien, K. 2001. "A Framework for the Establishment of Organizational Risk Indicators." *Reliability Engineering and System Safety* 147-167.

Ramesh, V., and R. Saravannan. 2011. "Reliability Assessment of a Co-Generation Power Plant in a Sugar Mill Using Fault Tree Analysis." *Energy Sources, Part A: Recovery, Utilization,and Environmental Effects* 33:12, 1168-1183.

Ribrant, Johan. 2006. *Reliability performance and maintenance, A survey of failures in windpower systems.* KTH School of Electrical Engineering, Master Thesis.

Saeed, Asif. 2008. *Online Condition Monitoring System for Wind Turbine Case Study.* Blekinge Institute of Technology.

Sterzinge, George. 2004. "*Wind turbine development, Renewable Energy Policy Project (REPP).*" Washington,.

Swain, A. D., and H. E. Guttmann. 1983. *Handbook of human reliability analysis with emphasis on nuclear power plant applications.* Washington, DC: US Nuclear Regulatory Commission.

Vasant, Pandian, Ivan Zelinka, and Gerhard-Wil Weber. 2019. "Intelligent Computing & Optimization." *ICO'2018, the 1st International Conference on Intelligent Computing and Optimization.* Springer International Publishing.

Vasant, Pandian, Ivan Zelinka, and Gerhard-Wil Weber. 2020. "Intelligent Computing and Optimization." *Proceedings of the 2nd International Conference on Intelligent Computing and Optimization 2019 (ICO 2019).* Springer International Publishing.

—. 2021. "Intelligent Computing and Optimization." *Proceedings of the 3rd International Conference on Intelligent Computing and Optimization 2020 (ICO 2020).* Springer International Publishing.

Woo, T. -H., and S. -H. Lee. 2012. "A Dynamical Study for Earthquakes of Seismic Safety Assessment in Nuclear Power Plants (NPPs) Using the System Dynamics (SD) Method." *Energy Sources, Part A: Recovery, Utilization, and Environmental Effects* 34:10, 948-958.

Zhang, Lin-lin, and Jie Li. 2007. "Probability density evolution analysis on dynamic response and reliability estimation of wind-excited transmission towers." *Wind and Structures, An Internationall Journal* 10 (1).

In: Advances of Machine Learning … ISBN: 978-1-68507-211-7
Editors: Pandian Vasant et al.

Chapter 3

Digital Technologies for the Implementation of Intelligent Diagnostics of the Insulation of Power Supply Systems with Insulated Neutral in Operating Mode

Svetlana Ovchukova[1], Nadezhda Kondrateva[2,*] and Andrey Shishov[2]
[1]Chuvash State Agricultural Academy, Cheboksary, Russia
[2]Izhevsk State Agricultural Academy, Izhevsk, Russia

Abstract

Nowadays, 75% of emergencies at power supply facilities are caused by single-phase ground faults. The currently used methods of thermal imaging and ultrasonic testing do not allow the accurate identification of the causes of accidents in electric power distribution systems. The current search generators, due to their low power, cannot cover the entire section

* Corresponding Author's Email: aep_isha@mail.ru.

of the diagnosed network. This article suggests the use of digital technologies for the intelligent diagnostics of the insulation of power supply systems with insulated neutral in the operating mode, allowing online diagnostics of the network and analyzing the data obtained for the early prevention of emergency situations. In order to avoid disconnecting sections and to diagnose the system while in operating mode, it is proposed to check the system using a current-limiting capacitor, which will be connected to the live parts of each phase under voltage and accurately determine the location of a single-phase short circuit for any number of feeders. The relevance of these studies is due to the fact that, for example, during two years at one of the manufacturing facilities, nineteen emergencies related to single-phase ground faults were recorded. Four of these cases led to a complete stop of production.

Therefore, the development of digital technologies for the implementation of smart diagnostics of the insulation of an energized power supply system with insulated neutral will be of interest to practitioners, scientists, researchers, graduate students and other specialists seeking to study the latest achievements in smart agribusiness for their subsequent application in the real world.

Keywords: insulated neutral system, insulation fault locations, hidden insulation faults, single-phase ground faults, WEB AI, ERP, automation

INTRODUCTION

To date, there is no known method in electric power distribution for diagnostics of branched power supply network insulation. Due to insulation failure, capacitive currents arise. Capacitive single-phase ground fault currents lead to the activation of the protector with the subsequent shutdown of the damaged section of the circuit. Further search for such damage therefore becomes more complex. The currently used methods of thermal imaging and ultrasonic testing do not accurately identify the cause of accidents in distribution networks. Furthermore, the current search generators, due to their low power, cannot cover the entire section of the diagnosed circuit. A phase-to-ground fault is abnormal and dangerous for a number of reasons:

- There is a danger of electric shock to service personnel and consumers
- Long-term presence of a single-phase ground fault in the network accelerates aging of insulation and causes damage to electrical equipment
- Ferroresonance phenomena can occur (especially with the alternate disconnection of power sources), from which voltage transformers most often fail [1].

MATERIALS AND METHODS

When calculating single-phase ground fault currents, a method based on the determination of the capacitive ground fault current is used. The values of the specific capacitive ground fault current can be found in the reference books (table 1) or taken from the technical characteristics of the equipment provided by the manufacturer.

Table 1. Specific value of capacitive currents in cable networks (A/km) [2]

Cables cross-section, $мм^2$	Capacitive current, I_C (A/km) at mains voltage	
	6кВ	10кВ
16	0,4	0,55
25	0,5	0,65
35	0,58	0,72
50	0,68	0,8
70	0,8	0,92
95	0,9	1,04
120	1	1,16

Table 1 does not show data on capacitive currents of overhead lines, generators and Busbar systems. Therefore, to determine the total capacity of the entire electrical installation, it is necessary to make a calculation.

The calculation of the specific capacitive current using the capacitance relative to the ground is carried out according to the formula below (1)

$$I_C = 3 \cdot U_F \cdot \omega \cdot C_O \cdot 10^{-3} \tag{1}$$

where:

$U_F = \frac{U_L}{\sqrt{3}}$ - Phase-to-neutral voltage (kV)

$\omega = 2\pi f = 314$ - Angular frequency (rad/s)

C_O - Phase-to-ground capacitance (μF/km)

The calculated values are used to set the relay protection settings for localizing the damaged area. The search for the source of the damage itself has to be carried out by field personnel. In order to avoid disconnecting the sections and to diagnose the network in operating mode, it is proposed to check the system using a current-limiting capacitor, which will be connected in phases to live parts under voltage. Thus, there will be a change in the phase capacity C_O relative to ground, as well as a zero shift, which will lead to an increase in the phase voltage to a linear value. The phase voltage will increase from $U_F = 3{,}7$ kV to values close to linear $U_F \approx 6{,}3$ kV.

This mode can be considered as a high-voltage test on equipment, which, according to regulatory documents for insulating equipment, is typically carried out only before installation into operation. If during the tests there is a loss of the insulating properties of the diagnosed object, the current modulator, which is connected in series with the current-limiting resistor, will be able to localize the place of the network fault. The method described above will allow:

- Accurate determination of a single-phase ground fault location for any number of feeders and their relative position in the vicinity of the switchgear.
- The search current does not spread over all outgoing feeders of the electrical installation, but flows directly through the place of the formed damage.
- The created search current allows diagnosing the electrical installation and detecting a malfunction in the stage of its formation.
- Allows finding cases of a self-healing fault occurrences and single-phase ground faults, which, with changing weather conditions, restores its insulating properties.

CONCLUSION

The obtained data can be integrated into a single ecosystem of smart use cases using the example of a "smart home" [3, 4]. Each large power supply facility, as a rule, has its own OPS servers, which collect and output data to the client's SCADA system. The measurement is executed after checking the insulation using the protocol RS-485 (Mudbus) or Ethernet (60870-104) can be collected in the local AWS system and transmitted via LTE | MQTT | HTTPS to a single WEB AI center, which in turn, according to the AI algorithm will monitor and analyze the power system. Data will be collected from hundreds of substations and thousands of consumer feeders integrated into a single system (Figure 1).

The described technological processes cover only the lower steps of the automation pyramid (Figure 1) and are limited to the level of Automated Process Control System (APCS) and Control & Instrumentation (CI). However, thanks to the collected and processed data from WEB AI, it is possible to establish and organize interaction with higher levels of the ERP system.

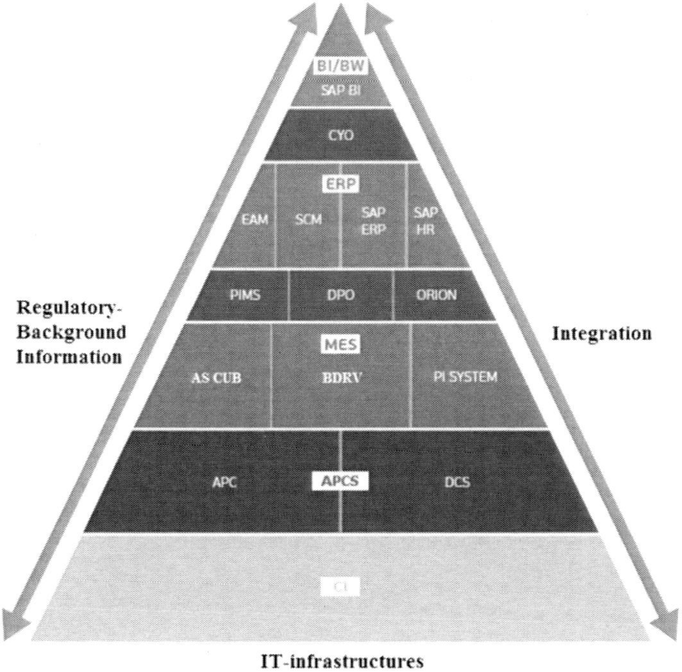

Figure 1. Pyramid of automation [5].

Future Research Works

To this date, the problem solved during the implementation of the project is significant not only for "smart" agriculture [6, 7], but also for industrial applications. In 2018-2019, 19 emergencies related to single-phase ground faults were registered at one of the gas production facilities. Four of these cases led to a complete shutdown of production. The existing methods of detection and localization of damaged sites are problematic and take a lot of time, since the inability to diagnose these damages in advance affects production [8]. The introduction of the proposed method into the production process will reduce the detection time and minimize losses. The tests of the proposed method for determining the locations of insulation damage were carried out at the "Pligon" substation in the city of Votkinsk. In all cases, the damaged area was localized. Thus, the development of digital technologies for the implementation of smart diagnostics of the isolation of an energized power supply system with insulated neutral is highly relevant and will be of interest to practitioners, scientists, researchers, graduate students and other specialists seeking to study the latest achievements in smart agribusiness for their subsequent application in the real world.

REFERENCES

[1] Gorunov V. A. Single phase ground fault. Is it possible to solve the problem? In the collection: *News of electrical engineering*. http://news.elteh.ru/arh/2017/104/04.php.

[2] Shabad M. A. *Calculations of relay protection and automation of distribution networks*. Moscow, 2018. 141-142 pp.

[3] Kondrateva N. P., Vashtiev V. K., Radikova A. V., Shishov A. A. Improving the reliability of operation of electrical equipment and 6-10 kV networks using digital technologies. *Topical issues of energy in the agro-industrial complex*. Materials of the National Scientific and Practical Conf. dedicated to the 100th anniversary of the GOERLO plan. Izhevsk State Agricultural Academy. Izhevsk, 2021. 42-45 pp.

[4] Kondrateva N. P., Filatov D. A. *Calculation of power lines for a greenhouse with LED irradiators*. Topical issues of energy in the agro-industrial complex. Materials of the National Scientific and Practical Conference dedicated to the 100th anniversary of the GOERLO plan. Izhevsk State Agricultural Academy. Izhevsk, 2021. 56-64 pp.

[5] Zorina S., *Automatic mode. Siberian oil*. Nov. 2013. http://www.uppro.ru/library/information_systems/management/avtomaticheski-regim.html.

[6] Kondrateva N. P., Filatov D. A., Terentyev P. V. *Justification of the energy-efficient length of power lines feeding greenhouses with LED*

irradiators. Bulletin of NGIEU: Nizhny Novgorod State Engineering and Economic University (Knyaginino), 2021. № 1. 54-63 pp.

[7] Kondrateva N. P., Terentyev P. V., Filatov D. A., Maksimov I. I., Kirillov N. K., Ovchukova S. A., Rybakov L. M. The effect of greenhouse irradiators on the load factor of step-down transformers. IOP Conf. Ser.: *Earth Environ. Sci.* 2020. 433 012039 https://doi.org/10.1088/1755-1315/433/1/012051.

[8] Kondrateva N. P., Filatov D. A., Terentyev P. V. Dependence of current harmonics of greenhouse irradiators on supply voltage. *Light&Engineering*. 2020. V.28. № 2. 85-88 pp. DOI: 10.33383/2019-024.

In: Advances of Machine Learning … ISBN: 978-1-68507-211-7
Editors: Pandian Vasant et al.

Chapter 4

IRRIGATION SYSTEM OF AGRICULTURAL FIELDS WITH THE USE OF SOLAR ENERGY

Leonid Yuferev and Alexander Parakhnich*
Federal Scientific Agroengineering Center VIM, Moscow, Russian Federation

ABSTRACT

This study analyzed the regions of the country where crops are being dyed in order to identify the possibility of using renewable sources to ensure the required irrigation rate. To calculate the power of the solar power plant, it was necessary to calculate the power consumption of the electric pump. The standard method for calculating and selecting an electric pump showed a large error. To improve the accuracy, a proprietary calculation method was developed based on operational tests of circulation pumps using machine learning using a regression algorithm. The new technique made it possible to reduce the error from 20% to 2–3%. Also, a schematic diagram of the operation of an autonomous irrigation system was designed and the choice of equipment was made.

* Corresponding Author's Email: leouf@yandex.ru.

Keywords: regression algorithm, irrigation, solar energy, pumps, renewable sources, calculation method

INTRODUCTION

At present, in the southern regions, in places remote from the central water and electricity supply, there is a problem of water supply for reclamation of agricultural crops. Russian agriculture is forced to spend much more energy and material resources for the production of products due to unfavorable natural and climatic conditions. The problem of water supply to regions with aridity coefficient of 0.11–1 covers an area of more than 250 million hectares, incl. 76 million hectares of arable land with great natural resource potential. They are located in 8 economic regions of Russia: Central, Central Black Earth, Volgo–Vyatka, Povolzhsky, North Caucasian, Ural, West Siberian, East Siberian. Of the total area of agricultural land in Russia (221,088.8 thousand hectares), more than three quarters (about 169 million hectares) are concentrated in arid territories. Of the total arable land in Russia (124373.8 thousand hectares), more than 80% (about 102 million hectares) is concentrated in arid territories. Of the total area of hayfields in Russia (23555.7 thousand hectares), 58% (more than 13 million hectares) is concentrated in arid areas. Of the total area of pastures in Russia (67367.7 thousand hectares), 73% (more than 49 million hectares) is concentrated in arid territories (Zonn et al., 2004). As a consequence, such areas require constant irrigation to obtain a larger crop. Irrigated agriculture, in contrast to rainfed, is a high–tech and highly costly production, but at the same time it is highly profitable, since irrigation can increase the productivity of agricultural crops by 2–3 times or more (Nikonov et al., 1987). It should be borne in mind that land reclamation will be effective only during the transition to a new stage of irrigated agriculture, where new knowledge, experience, technological means, new intensive varieties and hybrids, agrochemicals, etc. are required. The well–known solar water lifting units designed for use in agriculture powered by solar energy are divided into four groups:

- solar photovoltaic stations;
- the use of a thermodynamic Stirling engine in conjunction with an electric generator;
- solar turbines operating according to the Rankine thermodynamic cycle together with an electric generator;
- solar thermoelectric generators.

Investigation of solar thermoelectric generators and Stirling motors requires very high temperatures in the heating zone of hot junctions or operating heat, created due to the concentration of the light flux, hence, tracking systems for the apparent position of the Sun at sunrise. Solar photovoltaic batteries can operate both with and without concentration of the luminous flux when flat (planar) modules are installed permanently (Yuferev et al., 2019). In systems with a Stirling engine and a Ranken turbine, multiple types of energy are converted multiple times: solar radiation → heat → mechanical work → electricity, there is a complex kinematic scheme with rotary or reciprocating movement of elements, an additional electric machine is used an electromagnetic generator of direct or alternating current.

Solar photovoltaic and thermoelectric systems, being direct current sources, are characterized by a shorter energy conversion scheme: solar radiation (in a certain part of the spectrum) → electricity or the entire radiation spectrum → heat → electricity, respectively, they do not have rotating and moving parts, and, if necessary, obtain at the AC output are completed with a DC–to–AC converter (inverter) (Kharchenko et al., 2019).

At present, the most promising is the use of solar energy with its direct conversion into electrical energy by the photoelectric method (Kuznetsov et al., 2019, Kuznetsov et al., 2020). For power supply of autonomous water consumption objects, remote from power grids, by the method of generating electricity to drive the water–lifting mechanism, photovoltaic installations are the simplest and most reliable in real operating conditions, however, in terms of the unit cost of the generated electrical power and, therefore, in terms of the cost of the raised water, they still exceed solar ones, thermodynamic systems (Rejekiningrum et al., 2021, Chennnaif et al., 2020).

In this regard, it is proposed to use solar energy to extract water from deep sources. The advantages of using photovoltaic water–lifting installations are due to the seasonal coincidence of the maximum of the annual arrival of solar radiation and the annual maximum demand for water, with a relatively low probability of the presence of cloudy days in the spring–summer–autumn period (Metlov et al., 2016, Zyoud et al., 2020).

Another advantage of using a solar system as an energy source is that the frequency and volume of irrigation are correlated with solar activity, due to which such a photovoltaic water–lifting unit becomes self–regulating, efficient and reliable (Rubio–Aliaga et al., 2019).

It should also be noted that the development of water reclamation will contribute to solving a number of socio economic problems of the population of rural areas (Mukherjee et al., 2020).

Now there are many tables and maps of the potential of solar energy, a number of atlases are being developed with the territories of areas favorable for the construction of solar power plants, taking into account a number of natural, environmental, technical, economic, social factors, as well as maps of land resources; transport networks; specially protected territories; population residence card; power consumption dynamics map (Daus et al., 2016).

Calculation of enlarged indicators such as gross, technical and economic solar energy potential was carried out in a large number of works. However, this information allows us to identify only promising areas for the location of solar power plants, but it is insufficient to determine their composition, power and forecast their operating mode (Daus et al., 2018).

Purpose of the Study

Analyze the need for irrigation in the regions of the country, develop a method for calculating the power of an electric pump using machine learning using a regression algorithm, based on the data obtained, develop a block diagram of an irrigation system.

Materials and Methods

To determine the characteristics of the provision of the territory with moisture, the moisture coefficient K is used, which is determined by the formula:

$$K = \frac{P_0}{V_0}, \tag{1}$$

where P_0 – precipitation mm, V_0 – volatility mm.

So, for example, in Astrakhan the average annual precipitation is 234 mm, and the level of evaporation is 800 mm, substituting the values in formula 1, we get K = 0.29. Dry regions with a coefficient of evaporation (K <0.4) include Astrakhan region, part of the Volgograd region, part of the Samara region. Arid hot and warm southern regions (K = 0.40.7) include: part of the Samara region, Orenburg region, Rostov region, part of the Volgograd region, Saratov region, partly Krasnodar region, Stavropol region, Ufa region, part of Omsk and Novosibirsk region. Normally and not humidified (K = 0.71) include partly Krasnodar Territory, partly Stavropol Territory.

In regions with low rainfall and high volatility, high intensity solar energy, which can be used to extract water using photovoltaic plants from deep sources or rivers and lakes for irrigation of crops, and perennial plantings.

Irrigation of crops can be superficial or sprinkling and subsoil or drip. Surface irrigation by the nature of soil moistening and mechanization conditions is carried out by flooding along strips, areas or checks with flooding of the entire surface of the site (grasses, cereals) or with water supply along furrows (row crops). Sprinkling with moistening of the soil surface is carried out by sprinkling units (devices, wings with nozzles or trails) with spraying of water in motion or positionally, with water supply through pipes or with its intake from open sprinklers. With subsoil irrigation, the root layer is moistened (mainly due to the capillary rise of the Veda) from underground pipes with holes, porous pipes or mole holes, as well as by regulating the level of groundwater standing. Subsoil irrigation can also be used with double regulation of the water regime (irrigation and drainage). Drip irrigation is an irrigation method in which water is supplied directly to the root zone of the grown plants in controlled small portions using dropper dispensers. Allows to obtain significant savings in water and other resources (fertilizers, labor costs, energy and pipelines). Drip irrigation also provides other benefits (earlier harvest, prevention of soil erosion, less spread of diseases and weeds).

For the Republic of Crimea, temporary irrigation norms were adopted by the State Committee for Water Management and Land Reclamation in 2015 and are relevant now (Parahnich et al., 2020). The data are presented in Table 1 Temporary irrigation norms for determining the water demand limit in the years 50, 75 and 95% of water availability when growing crops on irrigation for the territory of the Republic of Crimea.

Table 1. Temporary irrigation rates for irrigation

Agricultural crop (crop group)	Irrigation rate, m^3/ha at			
	sprinkler irrigation with estimated provision,%			drip irrigation *
	50	75	95	
Spring planting potatoes	1300	1900	2400	1600
Summer planting potatoes	1300	1900	3800	12001500
Table grapes				13001800
Fruit orchards (apple tree)				14002000
Melon				1300
Watermelon				2100

The duration of the growing season lasts from the beginning of seed germination to the end of fruit ripening. This period of time will be different for different types of garden and horticultural crops. This concept means active growth of the plant, for example, for a cucumber from 90 to 110 days. If we talk about fruit trees, then the vegetative period of a plant of this type begins with the swelling of the buds in early spring, ends in the fall, after the foliage has fallen off. In winter, crops are in an inactive phase, so this period of time does not apply to the growing season. For example, the growing season of a summer apple is from 90 to 110 days (Parahnich et al., 2019). During this period, solar activity is also the highest.

Pumping units in agricultural production are used for pumping liquid media, viscous liquids and process, cooling and drinking water, they are also used for the extraction of water from deep sources. There are several types of pumps for pumping liquid in order to ensure irrigation: centrifugal, vibrating, axial, piston, screw, plunger, vortex (Pinto et al., 2020). Several types of pumps were considered and selected, suitable for the production of water from artesian wells, the data are summarized in Table 2

Table 2. Areas of application (selection) of pumps by capacity

<table>
<tr><th>To 10 m^3/hour</th><th>From 10 to 100 m^3/hour</th><th>From 100 to 1 000 m^3/hour</th><th>From 1 000 to 10 000 m^3/hour</th><th>From 10 000 m^3/hour and more</th></tr>
<tr><td colspan="3">Single stage centrifugal</td><td></td><td></td></tr>
<tr><td>Vibrating</td><td></td><td colspan="2"></td><td></td></tr>
<tr><td></td><td></td><td colspan="2">Multistage centrifugal</td><td></td></tr>
<tr><td colspan="2"></td><td colspan="3">Axial</td></tr>
<tr><td colspan="3">Reciprocating</td><td></td><td></td></tr>
<tr><td colspan="3">Screw</td><td></td><td></td></tr>
<tr><td>Plunger</td><td></td><td></td><td></td><td></td></tr>
<tr><td></td><td>Vortex</td><td></td><td></td><td></td></tr>
</table>

Centrifugal pumps are installed to draw water from wells in most cases. Such devices have a cylindrical body, inside which an electric drive is located. On the axis of the rotor of the motor, a turbine impeller is mounted, which operates in a chamber specially equipped for it. This turbine is made in the form of a snail; the device of the supply pipe is located in the center, and the outlet at the edges.

The accelerated rotation of the turbine contributes to the formation of centrifugal force, which throws the incoming water under pressure towards the outlet pipe. The result is a large difference in pressure readings. In the area closer to the shaft, a vacuum occurs, and the pressure increases along the edges of the blades. Thus, water constantly flows from the supply pipe during the rotation of the turbine, and the required water pressure is supplied at the outlet of the pump.

This principle of operation is used for most of the surface centrifugal pumps. But in such devices, the required output pressure indicators are often provided by increasing the size of the working chamber. In borehole pumps, the dimensions of the working chamber should be limited taking into account the diameter of the drilled hole. Therefore, the problem of increasing the pressure can be solved in another way. Several working chambers with turbines are installed on one axis. The water pressure pumped by the previous chambers is the support for each next one. As a result, the pressure indicator is formed from the total force of the installed impellers.

The number of chambers installed on a centrifugal well pump may differ depending on the depth to which it will be installed. Such a scheme fully justifies itself, since centrifugal pumps in the future are capable of taking water from a sufficiently large depth. Currently, the following series of centrifugal pumping units are produced: K, KM, ETsV, NMSh, AX, X, TsG, CNS, Grundfos, Wilo, Pedrollo, Calpeda, etc.

Centrifugal pumps are endowed with many advantages, thanks to which they are widely popular and in demand at the moment they are the most commonly used electric pumps for wells. The strengths of circulation pumps are summarized below:

Relatively high number of shaft revolutions. This makes it possible to use electric motors and turbines as mechanisms for ensuring rotation;

Possibility of gradual decrease or increase in power. Due to this quality, the startup of the device can be carried out with a closed valve at the outlet;

When several pumps are installed on the same pipeline, the pressure level and the intensity of the water supply increase significantly;

Simplicity of construction;
Comparative low cost;
Possibility of using an automated control system;
The liquid can be drawn into the pump from a great height;
The efficiency of such devices is approximately 0.60.8;
Reliability of operation.

Another used electric submersible pump is a vibration pump. The principle of its operation, which is based on the vibration of a moving armature with a membrane under the action of an electromagnetic coil. The membrane sucks in a portion of water through the check valve into the cavity of the mechanism and, during the return stroke, pushes it out of there through the outer fitting.

An axial flow pump is one of two types of dynamic vane pumps, the movement of the working fluid in which occurs in a continuous flow due to the interaction of this flow with the movable rotating blades of the rotor and the fixed blades of the housing. In this case, the portable movement of the working fluid flows in the axial direction, that is, parallel to the rotor rotation axis. They are mainly intended for conveying large volumes of liquid. Suitable for both liquids and gases, pumps for pumping gases are commonly referred to as axial compressors or axial fans.

The work of axial pumps is based on the force interaction of the blade with the flow around it. In axial flow pumps, the fluid flow is parallel to the axis of rotation of the impeller. An axialflow pump consists of a housing and an impeller that freely rotates in it. When the wheel rotates in the fluid flow, a pressure difference arises on both sides of each blade and, consequently, the force interaction of the flow with the impeller. The forces of pressure of the blades on the flow create a forced rotational and translational motion of the fluid, increasing its pressure and speed, that is, mechanical energy. The specific increase in the energy of the fluid flow in the impeller depends on the combination of flow rates, the speed of rotation of the wheel, its size and shape, that is, on the combination of design, size, speed and pump flow.

A piston pump (plunger pump) is one of the types of positive displacement hydraulic machines, in which the displacers are one or more pistons (plungers) that reciprocate. Unlike many other positive displacement pumps, piston pumps are not reversible, that is, they cannot operate as hydraulic motors due to the presence of a valve distribution system. Piston pumps should not be confused with rotary piston pumps, which include, for example, axial piston and radial piston pumps. The principle of operation is as follows: due to the translational movement of the piston, a vacuum is created in the cavity under it, and liquid is sucked there from the supply (suction) pipeline. When the piston moves back in the suction line, the valve is closed, preventing liquid from flowing back, and the valve on the discharge line, which was closed during suction, opens. The liquid that was under the piston is displaced there, and the process is repeated. The disadvantage of such a pump is that the liquid moves through the pipeline at different speeds (jumps). This point is usually bypassed by the creation of pumps in which several pistons. The main advantage is that it is able to pump liquid, being at the moment of startup unfilled with it (dry suction), and therefore it is usually used where this advantage must be taken advantage of. The displacement of the piston pump is proportional to the number of piston strokes and does not depend on the pressure. Currently, the following series of piston pumps are produced: НГС420, VVS/VVP, PVB, PVQ, PVH, PVXS, PVE, PVM, PVWS, PLATA, LVP, MVP, etc.

In screw pumps, the pressure of the pumped liquid is created due to the displacement of the medium by special screw rotors, which rotate synchronously in opposite directions inside the stator. The medium is pumped, moving along the axis of the screw in the working chamber, which is formed by the helical grooves and the inner part of the housing.

Screw assemblies provide the creation of a deep vacuum and are widely used in the food, chemical, pharmaceutical industry, in research institutions. They are actively used for freezing, drying, gas separation, vacuum filling and vacuum packaging, coating, etc.

The borehole screw pump is designed for lifting and pumping fresh water with a sand impurity content of not more than 150 g/m^3 from workings with a maximum depth of 50 m. The units are used for autonomous water supply of lowrise buildings, as well as for irrigation of household plots.

Vortex pumps are a type of vane pump. The impeller of vortex pumps is a massive steel disc with circumferentially milled grooves forming short rectilinear blades. The suction and discharge nozzles of the pump are usually located in the upper part of the casing, which ensures the subsequent selfpriming of the pump after a onetime priming at the initial startup of the pump. Inside the housing, concentric to the axis of the pump shaft, there is a discharge channel running in the direction of rotation from the inlet to the discharge branch pipe. A jumper is located between the suction and discharge nozzles, which fits to the impeller with a minimum clearance (0.150.2 mm) and overlaps at least 2 impeller blades. The jumper separates the suction cavity from the pressure chamber.

The principle of operation of vortex pumps, like centrifugal pumps, is based on the use of centrifugal force arising from the rotation of the impeller. However, there are some peculiarities in their work. When the pump impeller rotates, a certain amount of liquid from the suction pipe enters the grooves of the impeller and moves from the periphery to the center, that is, differently than in centrifugal pumps. Then this volume of liquid, under the influence of centrifugal force, begins to move along the blade, from the center to the periphery of the pump, and, having received velocity energy, is thrown into the ebb channel. In the channel, the velocity energy of the fluid volume is converted into pressure energy. Under the action of pressure and the suction action of the impeller blades, this volume of liquid again falls on the blades, and the cycle repeats. Thus, for a complete revolution of the impeller, this cycle is repeated many times, and each time there is an increase in energy and, consequently, pressure.

Discussion Results

After calculating the required amount of water per day at the upper rate for drip irrigation of orchards $\text{К}_y = 2000 \text{ m}^3/\text{ha}$ for the growing season $\text{В}_p = 110$ days, we get the following expression

$$G = \frac{\text{К}_y}{\text{В}_p} = \frac{2000}{110} = 18{,}18 \ \frac{m^3}{ha*days}, \tag{2}$$

Let's calculate the required amount of water for irrigation per hour with an average effective duration of a sunny day of 6 hours, for systems without tracking the position of the sun:

$$G_h = \frac{18{,}18}{6} \approx 3 \ \frac{m^3}{ha*days}. \tag{3}$$

The calculation was based on the standard procedure for choosing a cantilever circulating electric pump.

Power calculation according to the standard method includes: determination of the coefficient of speed

$$n_s = \frac{3.65*n*\sqrt{Q}}{H^{\frac{3}{4}}}, \tag{4}$$

determine the volumetric efficiency

$$\eta_o = \frac{1}{1+0.68*{n_s}^{-0.66}}, \tag{5}$$

calculate the reduced diameter

$$D_{1n} = 4.25\sqrt[3]{\frac{Q}{n}}, \tag{6}$$

calculate the hydraulic efficiency

$$\eta_g = 1 - \frac{0.42}{(lg\ D_{1n} - 0.172)^2}, \tag{7}$$

the mechanical efficiency is taken as a constant value $\eta_m = 0.93$ find the full efficiency

$$\eta = \eta_o \eta_g \eta_m, \tag{8}$$

formula for calculating the power of the electric pump

$$P = \frac{\rho * g * Q * H}{1000 * \eta}, \tag{9}$$

According to the production tests of the Calpeda 4SDF 46/21EC pump, the theoretical power was calculated according to the formula 9, for clarity of the results, all data are summarized in Table 3.

For clarity, the result obtained from Table 1 is presented in the form of graphs in Figure 1.

Table 3. Data for checking the accuracy of the calculation method

Calpeda 4SDF 46/21EC			
Consumption, m³/hour	Pressure, m	Factory Power Tests, kW	Theoretical power formula 9
0,90	117,00	1,09	**0,59**
1,20	114,00	1,16	**0,72**
1,80	108,00	1,26	**0,95**
2,40	100,00	1,36	**1,12**
3,00	91,90	1,42	**1,25**
3,60	82,20	1,47	**1,30**
4,20	71,40	1,47	**1,29**
4,80	59,60	1,41	**1,20**
6,00	32,10	1,30	**0,77**

Curve 1 graphically displays the results of factory tests, curve 2 displays the theoretical calculation data.

Figure 1 shows that the theoretical calculation according to the known technique has a large error in comparison with the tests of a real pump. To obtain a more accurate value of the required power of the electric pump, a calculation method was developed based on a machine learning algorithm. Figure 2 below shows the learning algorithm.

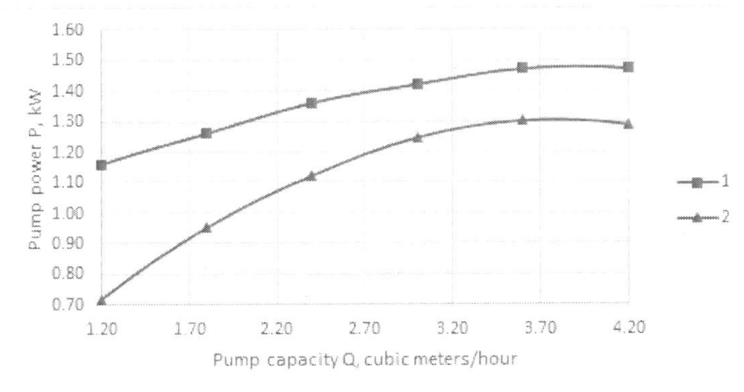

Figure 1. Dependence of power characteristics on pump performance Calpeda 4SDF 46/21EC.

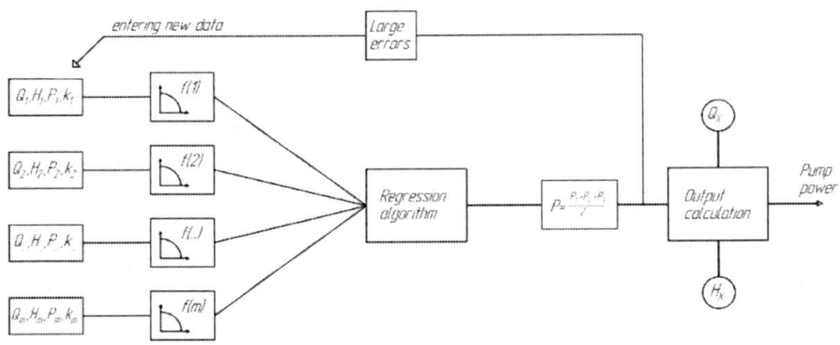

Figure 2. Algorithm for calculating power based on regression analysis.

Regression algorithms can predict the value of a new data point based on previously acquired data (Kropat et al., 2018, Ozmen et al., 2014). For the training, the data of production tests of centrifugal submersible pumps of the following manufacturers were used: Calpeda SDF series, Pedrollo 4SR series, Ebara, etc. To plot the characteristics of the pumps, a capacity range from 30 to 70 liters/min was chosen, which corresponds to 1.8 to 4, 2 m^3/h, in this range there were pumps with a head from 90 to 396 m, the power of electric pumps in a given range was from 2.2 to 5.5 kW.

The learning algorithm takes place in several stages: first of all, a selection of pumps in the required range is made, in our case, pumps were selected that can provide water, the previously calculated flow rate for a summer apple tree. At the second stage, according to the data obtained for each pump, graphs of dependencies are plotted, shown in Figure 3, at the third stage, a regression analysis of the obtained dependencies is performed, as a result of this, the characteristic of all pumps in the X, Y, Z planes is built using the Curve Fitting algorithm, as a result of which we obtain a regression function formula 10, a graphic representation of which is shown in Figure 4.

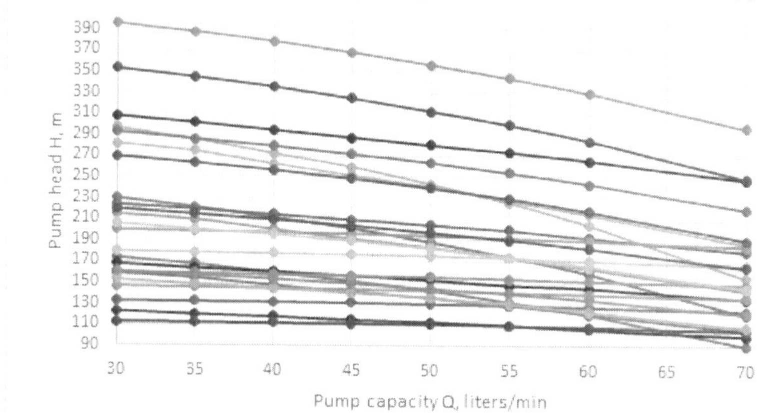

Figure 3. Graphical characteristics of factory tests of pumps.

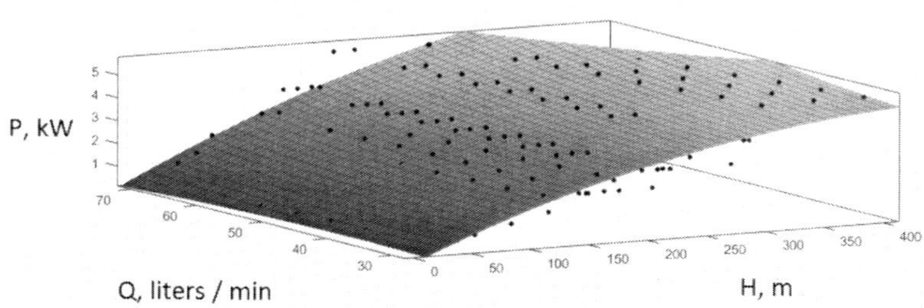

Figure 4. Graphical representation of the regression function of the dependence of the power (P) of the pumps on the depth (H) and capacity (Q).

$$P_1 = 0.019 \cdot \mathrm{H} - 0.033 \cdot \frac{\mathrm{Q}}{0.066} - 2.179 \cdot \mathrm{H}^2 \cdot 10^{-5} + 3.267 \cdot \mathrm{H} \cdot \frac{\mathrm{Q}}{0.066} \cdot 10^{-5} + 41.06 \cdot \frac{\mathrm{Q}}{0.066}^2 10^{-5}. \quad (10)$$

At the fourth stage, "activation function", the data of the regression function number 10 is processed, followed by the summation of the results of power calculations according to the following formula 11. As a result, it is possible to determine the power consumption of any pump in the capacity range from 30 to 70 liters/min, which corresponds to 1, 8 to 4.2 m^3/h, in this range there were pumps with a head from 90 to 396 m.

$$P = \frac{P_1 + P_2 + P_3}{Z}, \quad (11)$$

where P_2, P_3 are the conversion factors for the power of the electric pump; Z factor of proportionality within 3.6.

$$P_2 = 0{,}4984 * \ln Q + 2.0332, \quad (12)$$

$$P_3 = \frac{\rho * g * Q * H}{1000 * \eta * 3600}, \quad (13)$$

At stage five "feedback", the obtained result of the power calculation is compared with the values of the factory tests of the pumps. In the case of a large error, "teacher's help" and the input of additional data from production tests of pumps, which were not previously included in the database for machine learning and to reduce the error in calculating the power consumption, are required.

At the sixth stage "Output calculation", the power consumption of the electric pump is calculated according to the given parameters of the flow rate Q and the head H. To check the reliability of the received machine melody of the calculation, the data of the Calpeda 4SDF 46/21EC pump were used.

According to the production tests of the Calpeda 4SDF 46/21EC pump, the theoretical power was calculated according to the formula 10 and the calculation was made according to the formula 4, for clarity of the results, all data are summarized in Table 4.

For clarity, the result obtained from Table 3 is shown graphically in Figure 5.

Curve 1 graphically displays the results of factory tests, curve 2 displays the theoretical calculation data, curve 3 displays the calculation based on the regression algorithm.

Table 4. Data for checking the accuracy of the calculation method based on the regression algorithm

Calpeda 4SDF 46/21EC				
Consumption, m^3/hour	Pressure, m	Factory Power Tests, kW	Theoretical power formula 9	Power parameters formula 11
0,90	117,00	1,09	0,59	1,18
1,20	114,00	1,16	0,72	1,23
1,80	108,00	1,26	0,95	1,30
2,40	100,00	1,36	1,12	1,35
3,00	91,90	1,42	1,25	1,40
3,60	82,20	1,47	1,30	1,44
4,20	71,40	1,47	1,29	1,47
4,80	59,60	1,41	1,20	1,49
6,00	32,10	1,30	0,77	1,49

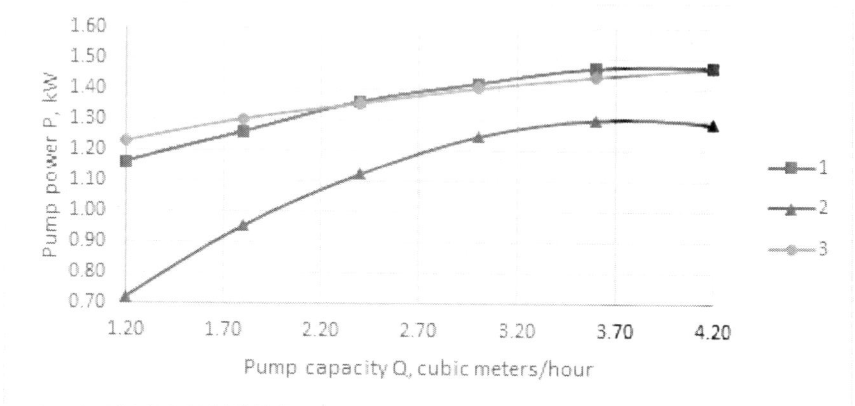

Figure 5. Dependence of power characteristics on pump performance Calpeda 4SDF 46/21EC

The method was also checked with the search for the power of the Calpeda 4SDF 54/10EC pump with other parameters, which are presented in Table 5.

Comparison of the calculation results for the Calpeda 4SDF 54/10EC pump is shown in Figure 6.

Table 5. Data from production tests and calculation based on the regression algorithm

Calpeda 4SDF 54/10EC				
Consumption, m^3/hour	Pressure, m	Factory Power Tests, kW	Theoretical power formula 9	Power parameters formula 11
0,60	94,80	0,76	0,34	0,91
1,20	89,70	0,90	0,55	0,99
1,80	83,40	0,99	0,72	1,04
2,40	75,30	1,03	0,83	1,06
3,00	65,70	1,06	0,87	1,08
3,60	54,60	1,04	0,85	1,08
4,20	42,60	1,03	0,75	1,07
4,80	30,00	0,94	0,59	1,06

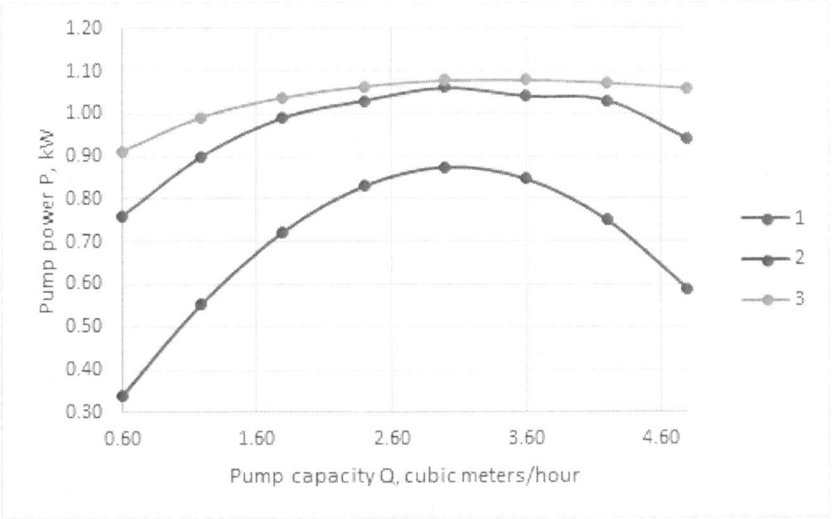

Figure 6. Dependence of power characteristics on pump performance Calpeda 4SDF 54/10EC.

Curve 1 graphically displays the results of factory tests, curve 2 displays the theoretical calculation data, curve 3 displays the calculation based on the regression algorithm.

From the above, it is fashionable to conclude that the theoretical calculation method had a very large error of the order of 20% at the extremum point in comparison with factory tests. After processing the data using a regression algorithm, the error was reduced to 23%, which is enough to calculate the power consumption of the submersible circulation pump, the data of which will be used to calculate the solar station.

Next, we will calculate a solar station for the extraction of water from deep sources in order to provide water supply to the irrigation system. For example, the depth of the aquifer is H = 100 m, the required productivity is $G_{hour} = 3$ m^3/(ha·hour). According to the formula number 11, we get the power consumption of the pump will be P $\approx$ 1.5 kW, which just corresponds to the nominal for Calpeda 4SDF 46/21EC. Consequently, with an average effective duration of a sunny day of 6 hours, the power consumption of the pump will be 9 kW·h/day.

To determine the generated power by a solar panel, we use the following formula

$$E = \frac{I * P_S}{U * K_O * K_P}, \tag{14}$$

where E is the real power of the battery kW·h; I solar radiation kW·h/m²; P_S = 250W is the power rating of one solar cell 250M JT250SCc; U is the amount of solar radiation that the manufacturer was guided by when calculating the rated power of 1000 W/m²; K_O = 1.11 is the correction factor for the amount of solar energy falling on the panel. It depends on the angle of inclination of the battery and the angle of its deviation from the south direction; K_P = 0.85 is a coefficient that characterizes how much electrical energy is lost in the entire autonomous power supply system with proper assembly and good equipment is about 15%. To correctly determine the power generated by the solar panel, we will use Table 6 and calculate the average level of solar radiation in the period from April to September I_{AS}, since watering is required during this period (Zvyagintsev et al., 2015).

Table 6. The daily amount of solar radiation in Simferopol is

Month	Solar radiation, kW·h/m2	Optimal tilt angle, degrees
January	2,26	62
February	2,78	53
March	3,65	39
April	4,58	24
May	5,51	14
June	5,89	6
July	6,28	9
August	5,62	21
September	4,75	36
October	3,81	51
November	2,65	60
December	1,86	63
Average for the year	4,14	36

$$I_{AS} = \sum_{S}^{A} \frac{I}{n} = \frac{4{,}58+5{,}51+5{,}89+6{,}28+5{,}62+4{,}75}{6} = 5{,}44\ KW * \frac{h}{m^2} \quad (15)$$

Since it was calculated earlier the power consumption of the pump will be 9 kW·h/day, therefore, about 7 photovoltaic panels 250M JT250SCc will be required to provide the pump with electricity.

DESCRIPTION OF THE DEVELOPED IRRIGATION SYSTEM

The principle of operation of the irrigation system using solar energy.

Photovoltaic station 6 is a primary source of energy capable of converting light energy into electrical energy. The PV plant for this case was previously calculated based on the pump power, the station includes 7 250M JT250SCc PV modules. The photovoltaic modules are connected in series with a total voltage of 223 V. Module Efficiency 15.2%; Dimension 1655x7000x40 mm.

Since the use of storage batteries is not advisable due to the high cost and low service life, it was decided to replace them with a block of capacitors 7 necessary to compensate for the starting currents of the electric pump and control the solenoid valve 4. Due to the fact that the accumulation of water occurs in the storage tank 3 water accumulation occurs only in sunny weather. As a result, there is no need to accumulate electrical energy in the battery station. Since the cost of batteries in standalone systems is about 60% of capital costs, abandoning them will significantly reduce capital costs. If irrigation is necessary and there is no sufficient level of solar radiation for the pump to operate, a hydroaccumulator tank with a capacity of 32 m^3 is used, this volume should be enough for 34 days to ensure irrigation in cloudy weather.

The control unit 8, based on the data received from the water level sensor in the well, determines the possibility of starting the electric pump 2. Also, the control unit 8, based on the data received from the float water level sensor in the accumulator tank 3, makes a decision to turn off or start the electric pump 2 located in the well 1 at a depth of 3 meters from the dynamic water level. During the nonuse of the irrigation system, the pumping station will raise water as long as there is solar radiation in the hydroaccumulator tank until the latter is full. This will make it possible to economically use underground water sources, and to use almost 100% of the energy obtained from solar panels.

Based on the data received from the soil moisture and precipitation sensor 10, the control unit decides on the need for irrigation and controls the solenoid valve 4. During natural precipitation, using soil moisture and precipitation sensors, the system, based on the data received, will decide what at a given moment in time there is no need for watering. After precipitation falls, the system will analyze the rate of soil drying and, from the calculation of irrigation standards, will include irrigation in order to perform it, while the amount of water that has fallen in the form of precipitation will be subtracted from the required amount of water. Also, on the basis of data on the vegetative stage of plant development, the required irrigation dose is normalized.

Irrigation network 5 consists of a pipeline and drip irrigation canals of the required length. The use of drip irrigation will allow you to regulate the water supply by adjusting the flow rate using the electro valve 4 and will help save water resources. The supply of water in portions throughout the entire period of plant development will maintain the required level of soil moisture. Another advantage of drip irrigation is the ability to mix fertilizers that go directly to the root system area using a separate container or directly in a hydroaccumulator tank.

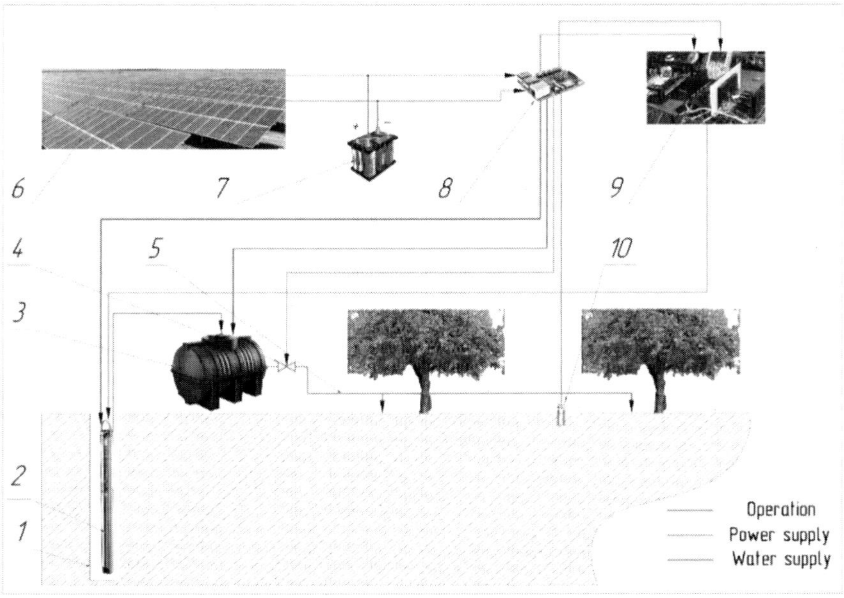

1 borehole; 2 submersible circulation pump; 3 hydroaccumulator; 4 solenoid valve; 5 irrigation network; 6 photovoltaic station; 7 intermediate block of capacitors; 8 control unit; 9 frequency converter (inverter); 10 sensor of soil moisture and precipitation.

Figure 7. Schematic diagram of a solar powered irrigation system.

Frequency converter (inverter) 9 is a device for converting direct current into alternating current with a change in voltage value. Usually it is a generator of a periodic voltage, in the form of an approximate sinusoid, or a discrete signal. The inverter turns on at a time when there is a need for the pump and there is sufficient solar radiation.

The accumulator tank should be located at the highest point of the field to create the necessary pressure in the irrigation system. This technical solution will make it possible to do without an additional electric pump, since water will flow through the irrigation system spontaneously, only the electric valve 4 will regulate the flow.

CONCLUSION

The conducted analytical review showed that the problem of irrigation covers 8 economic regions of Russia: Central, Central Black Earth, VolgoVyatka, Povolzhsky, North Caucasian, Ural, West Siberian, East Siberian. The use of renewable energy sources in these areas is due to the high solar activity in the summer. This allows the use of photovoltaic power plants as a primary source of energy for the extraction of water from underground sources. To design an autonomous photovoltaic station for the purpose of irrigation, it is necessary to know the irrigation rate, the depth of the aquifer, the power of the electric pump, and the daily amount of solar radiation in the region. For the extraction of water from underground sources, a pump analysis was carried out, which showed that circulation pumps have a number of advantages. The main advantages are the possibility of parallel operation, frequency control with the use of automation, a wide range of models, a fairly high efficiency, etc. The processing of data on the characteristics of factory tests of circulation pumps using a regression algorithm made it possible to develop a simpler and more accurate method for calculating power. Which has an error of the order of 23% compared to the standard technique, the error of which was 20%. On the basis of data for the Republic of Crimea, the daily rate of drip irrigation of a summer apple tree was calculated. Also calculated the actual generated power per day by the photovoltaic panel 250M JT250SCc for the city of Simferopol for the period from April to September. A schematic diagram of the operation of an irrigation system using photo electricity was developed equipment was calculated and selected.

REFERENCES

Chennaif, M., Elhafyani, M. L., Zahboune, H. & Zouggar, S. (2020). Electric system cascade analysis for optimal sizing of an autonomous photovoltaic water pumping system. *Lecture Notes in Electrical Engineering, 282290.*

Daus, Y., Kharchenko, V. & Yudaev, I. V. (2018). Solar radiation intensity data as basis for predicting functioning modes of solar power plants. *Handbook of Research on Renewable Energy and Electric Resources for Sustainable Rural Development*, 283–309.

Daus, Y. V., Kharchenko, V. V. & Yudaev, I. V. (2016). Evaluation of solar radiation intensity for the territory of the Southern Federal District of Russia when designing microgrids based on renewable energy sources. *Applied Solar Energy (English translation of Geliotekhnika)*, *52*(2), 151–156.

Kharchenko, V., Nikitin, B., Tikhonov, P., Panchenko, V. & Vasant, P. (2019). Evaluation of the Silicon Solar Cell Parameters. *Advances in Intelligent Systems and Computing*, *866*, 328–336.

Kharchenko, V. & Vasant, P. (2018). Handbook of research on renewable energy and electric resources for sustainable rural development. *Handbook of Research on Renewable Energy and Electric Resources for Sustainable Rural Development*, *1672*.

Kropat, E. & Weber, G. W. (2018). Fuzzy targetenvironment networks and fuzzyregression approaches. *Numerical Algebra, Control and Optimization*, *8*(2), 135155.

Kuznetsov, P. N., Abd Ali, L. M., Kuvshinov, V. V., Issa, H. A., Mohammed, H. J. & AlBairmani, A. G. (2020). Investigation of the losses of photovoltaic solar systems during operation under partial shading. *Journal of Applied Engineering Science*, *18*(3), 313330.

Kuznetsov, P. N., Lyamina, N. V. & Yuferev, L. Y. (2019). A device for remote monitoring of solar power plant parameters. *Applied Solar Energy (English translation of Geliotekhnika)*, *55*(4), 247251

Kuznetsov, P. & Yuferev, L. (2019). Feature of operation PV installations with parallel and mixed commutation photocells. *Advances in Intelligent Systems and Computing*, *866*, 231238.

Metlov, G. N., Korolev, V. A. & Belenov, A. T. (2016). Mobile solar photovoltaic waterlifting installations for irrigation. *Bulletin of Agrarian Science of the Don*, *4955*.

Mukherjee, P. & Sengupta, T. K. (2020). Design and Fabrication of SolarPowered Water Pumping Unit for Irrigation System. *Computational Advancement in Communication Circuits and Systems*, *89102*.

Nikonov, A. A. (1987). Scientific bases of sustainable development of agricultural production in arid regions of the country. *Bulletin of agricultural science*, *1025*.

Ozmen, A., Kropat, E. & Weber, G. W. (2014). Spline regression models for complex multimodal regulatory networks. *Optimization Methods and Software*, *29*(3), 515534.

Parakhnich, A. S. & Yuferev, L. Y. (2020). Preliminary calculation of the parameters of a solar waterlifting installation. *Electrotechnology and electrical equipment in the agroindustrial complex*, *132137*.

Parakhnich, A. S., Yuferev, L. Y. & Gusarov, V. A. (2019). The use of photoelectric waterlifting installations for irrigation in the Republic of Crimea. *Innovations in agriculture*, *8593*.

Pinto, R., Mathias, C., Kokande, N. & Thomas, M. (2020). Solar powered irrigation system. *Nanoelectronics. Circuits and Communication Systems*, *369381*.

Rejekiningrum, P., Apriyana, Y. & Harmanto. (2021). The application of solar water pump for drip irrigation to increase shallot yield on dry land. *IOP Conf. Series: Earth and Environmental Science*, doi:10.1088/17551315/648/1/012091.

RubioAliaga, A., GarciaCascales, M. S., SanchezLozano, J. M. & MolinaGarcia, A. (2019). Multidimensional analysis of groundwater pumping for irrigation purposes: Economic, energy and environmental characterization for PV power plant integration. *Renewable Energy*, *174186*.

Yuferev, L. & Kuznetsov, P. N. (2019). Increasing the energy efficiency of photovoltaic systems operating under conditions of uneven illumination. *In Renewable Energy and Power Supply Challenges for Rural Regions, 88114.*

Zonn, I. S., Trofimov, I. A., Shamsutdinov, Z. S. & Shamsutdinov, N. Z. (2004). Land resources of arid territories of Russia. *Arid ecosystems, 87101.*

Zvyagintsev, Y. M. & Yuferev, L. Y. (2015). The system of autonomous power supply of the farm on the territory of the Autonomous Republic of Crimea. *Innovations in agriculture, 100103.*

Zyoud, A., Othman, A., Manasrah, A. & Abdelhafez, E. (2020). Investment analysis of a solar water pumping system in rural areas in Jordan. *International Journal of Energy for a Clean Environment, 269281.*

In: Advances of Machine Learning … ISBN: 978-1-68507-211-7
Editors: Pandian Vasant et al.

Chapter 5

STRATEGIES HYBRID SIMULATION FOR REGIONAL MARKET DEVELOPMENT OF RENEWABLE ENERGY

P. N. Kuznetsov*[1,*], *PhD, D. Yu. Voronin*[1], *PhD, L. Yu Yuferev*[2], *DSc and V. P. Evstigneev*[1], *PhD
[1]Sevastopol State University, Sevastopol, Russian Federation
[2]Research Agricultural-Engineering Center VIM, Moscow, Russian Federation

ABSTRACT

One of the main problems of the energy industry organization is the constant need to find compromise between the demand for electricity and its available volume at the current time. The volatility of the electricity generation using renewable energy sources compound the considering scientific problem. Decision-making procedures of operational dispatch control require special attention. This work is devoted to the development of effective tools for countering the a priori uncertainty of the demand and supply dynamics for the distributed electricity market with adaptive procedures for finding compromise solutions for the proactive

[*] Corresponding Author's Email: pnkuznetsov@sevsu.ru.

management of the regional electricity market. The proposed and implemented information technology bases on AnyLogic multi-approach simulation environment aimed to support decision-making on the choice of the optimal strategy for the development of the renewable energy market in the region. It focuse on organizing an energy complex built on the principles of distributed generation using renewable and conventional power plants to supply the region with energy. The practical use of this technology will reduce the likelihood of a monopoly on the energy market of the region, which poses a particular danger to its economy. The results of numerical modelling carried out that the using of the proposed approach for most regions is incorrect. Share of power generating plants using renewable energy resources should be about 25% for regions with a share of industrial consumers constituting less than 40%, and for other regions no more than 16%.

Keywords: energy market, renewable energy, hybrid simulation, decision-making strategies

INTRODUCTION

At present, the energy stability of the world in general, and of each of its individual regions in particular, is the foundation of national security and a fundamental factor in development, prosperity and social growth. The use of our own environmentally friendly sources of energy allows us to solve not only the energy supply of the state, but also to ensure the fulfillment of international obligations (Borenstein et al., 1999; Belyaev et al., 2002).

The stability of the electric power industry depends on a number of factors, among which the degree of uniformity of generation, as well as the development of reserves for emergency provision of peak loads in the energy sector, are essential (Voropai et al., 2004).

Thermal power plants still make up the bulk of the electric power industry in most countries, which is explained by technological processes and possible modes of operation in various situations in the power system: both reliable guaranteed supply and provision of peak loads when necessary are possible. Unlike thermal power plants, nuclear power plants are limited in terms of the flexibility of the electricity supply schedule, and the creation of hydroelectric power plants requires territorial losses in the agricultural sector, the presence of large river resources, as well as the corresponding features of the relief. The last two options for energy production sources require significant costs for their construction, and in the case of nuclear power - for the disposal of radioactive equipment and structures at the end of their service life (Dittmar et al., 2012).

The last century is connected with an active growth in research and implementation of alternative solutions to reduce the cost of generating energy, improve the environmental friendliness of regions and increase the reliability and availability of the system. Such a transition implies an increase in the use of distributed generation, which requires intelligent decentralized control, ensuring the free exchange of electricity between consumers, generating companies and prosumers (Bogachkova et al., 2018; Ines et al., 2020). The use of renewable energy sources provides not only new opportunities for achieving prosperity and well-being of regions and countries, but also the possibility of promising economic and technological development of the territory where it is impractical to use traditional energy or there are no necessary resources for their reliable operation.

Effective introduction of power plants with unstable generation into the power system requires the creation of developed intelligent technologies for the distribution and dispatching of electricity. Such technologies should ensure the reliability and availability of energy supply, as well as comply with the basic principles of sustainable development (environmental, economic and social). In particular, for the effective implementation of the technology of intelligent active-adaptive networks "SmartGrid", it is necessary to create developed technologies for predicting energy consumption in the micro- and mesoscale (Ahmad et al., 2020; Kapitonov, 2019).

The above tendencies necessitate the development and implementation of methods for forecasting and dispatching energy consumption, contributing to the sustainable development of the regional energy system. These circumstances aim to give solutions that correspond to the basic principles of sustainable development of regions, imply the use of modern end-to-end information technologies to increase the efficiency of the use of natural resources and improve the technical and economic indicators of the electric power industry.

AnyLogic is a multi-approach simulation environment to support decision-making. The choice of the optimal strategy for the development of renewable energy market in the region was made using AnyLogic. The proposed and implemented information technology bases on its functional opportunities. It focuses on organizing an energy complex built on the principles of distributed generation using renewable and conventional power plants to supply the region with energy. The proposed methodology integrates methods of network analysis of projects, agent-based and discrete-event paradigm for the synthesis of simulation models of complex systems, which makes it possible to increase the efficiency of decisions on organizing the energy market in the region.

The practical use of this technology will reduce the likelihood of monopoly on the regional energy market, which poses a particular danger not only for the regional economy, but also for the state. In other words, it will make it possible to determine the optimal strategy for distributing the share of participants in the electricity market. It will hinder the formation of the so-called “supplier market power”, which consists in the ability of one of the participants to unilaterally influence the formation of the cost of electricity, providing its own excess profitability compared with a competitive situation in the market (Wadhwa, 1989). As a result, the development of information technology to support decision-making on reengineering and regulation of the electricity market is an extremely important task.

ANALYTICAL FRAMEWORK

Literature Review

To solve effectively the scientific problem under consideration, it is required to conduct a deep analysis of existing approaches to assessing, visualizing and managing the energy balance in the region. The main subject areas of obtained scientific results are the following:

1. Synthesis of a unified intellectual methodology for supporting decision-making on energy balance management in the region. The proposed approach has a complex hybrid structure and a multilevel system for the distribution and generation of electrical energy. The works (Ma et al., 2010; Samakpong et al., 2019) describe architectural solutions for information systems for dispatching power systems, principles of integration with external sources. In (Ma et al., 2010), a stream computing model with a dynamically generated power consumption request is described, and the features of its architectural implementation when using supercomputer computing are disclosed. In (Fu et al., 2015) multi-agent methods and tools for efficient energy supply management are considered. In (Roche et al., 2010; Mitiku et al., 2020), a system for modeling cloud services is presented, which makes it possible to provide energy companies with data on the feasibility of reengineering the processes of distribution and generation of electricity in accordance with the predicted or real values of the analyzed parameters. The existing methods in most cases base on analytical models of the theory of mass service and adaptive optimization algorithms of a heuristic nature (Li et al., 2018; Alhendawi et al., 2019).
2. Using the mathematical apparatus of game theory to find the optimal strategy for finding compromise solutions for the development of the energy market in the region. The work (Vasil'ev, 2016) proposes the specification of Berge's "altruistic equilibrium", taking into account the combination of altruism, economic egoism and the influence of external factors on the preferences of economic agents. In (Geraskin, 2017), the expediency of using the apparatus of the theory of reflexive games to search for equilibrium controls when making decisions in ecological-economic systems is discussed.

3. Development of ideas of dual control in technical systems. The proposed approach (Geraskin, 2003) use nonparametric dual control algorithms to solve problems of identification and management of the development of the electric power industry in the region under conditions of incomplete information.

When solving the problems of efficient organization of the energy system of the region, it is important to perform the decomposition of the considered tasks. The classical approach of systems theory, unfortunately, has low constructiveness and leads to very limited results. This is primarily because the elements of the energy system form a single structure in accordance with some target feature, external and internal factors (Geraskin, 2003, Tao et al., 2019). They can be economic in nature, driven by the energy needs of the region and meet environmental and functional constraints. In a number of cases, their own peculiarities appear. System-forming factors certainly exist, but the use of centralized concepts to describe scenarios for the functioning of energy systems has a very weak constructiveness. An alternative description is the reference model proposed in (Geraskin, 2003). Further, the following basic Actors (Yaghooti et al., 2018) of the reference model are considered: Consumer, Generator and Distributor, as well as extended (Yaghooti et al., 2018): Dispatcher, Controller. The works (Yaghooti et al., 2018; Lospinoso et al., 2011) describe their functional features and basic interaction scenarios.

For the energy system of the region, we will single out the following states: safe, efficient, pre-critical and emergency. Safe states characterize high stability of the system, that is, the absence of the possibility of a direct transition to pre-critical states. With insufficient efficiency of procedures for ensuring guaranteed operability, the power system tends to degradation, that is, a transition from operable states to states close to critical. In this case, if the stabilizing processes of its reconfiguration launch in a timely manner, then even a minor failure can lead to an unrecoverable failure and a transition to an emergency (absorbing) state. In accordance with (Tan et al., 2021), it is customary to distinguish two basic approaches to solving the problem of managing the development of the electric power industry in the region: proactive and reactive. With proactive management, the main goal of the decisions made is to prevent the occurrence of various adverse events. This approach focuses on the use of predictive assessments of the development of risks and describes the latent part of their life cycle in conditions of a lack of information on the magnitude of the possible consequences of the development of negative events. The reactive approach characterize the presence of accurate assessments of the consequences of incidents that have occurred, and the main goal is to neutralize the negative consequences of events that have already occurred. Figure 1 shows the key features of proactive and reactive management of the electric power industry in the region in more detail. The proactive approach focuses mainly on the use of preventive measures and is effective for both highly complex and relatively dynamic systems. On the other hand, control actions in the reactive approach are more costly than in the proactive approach. The reactive approach allows you to save on monitoring, but at the same time, the likelihood of getting into an unfavorable state becomes significantly higher. It is known (Elluru et al., 2019, Yuvaraj et al., 2019), that the purpose of reactive control is to prevent the transition of the object of critical application to states with an "unacceptable" level of system functioning, and proactive - to states having an "essential" level. Proactive and reactive approaches are more complementary than competing. Each of them has its own area of effective application and, in some cases, only the Dispatcher can make the best choice in favor of using one of them.

In the classical approach, the elements of the analyzed system are endowed with clearly expressed functions focused on achieving a certain general goal, and it is in this connection that their usefulness, necessity, effect, and degree of development are assessed (Clayton et al., 2018). Each of the basic Actors has its own relative nature of behavior, its own goals, and the quality of the organization of the energy complex of the region. The functioning of the regional electric power market focuses on achieving one (or many) goals. In accordance with the reference model (Clayton et al., 2018; Monteiro et al., 2016; Kuznetsov et al., 2019; Yuferev et al., 2019), the process of functioning of the electricity market can be described by the interaction of basic Actors having, in accordance with their type (Consumer, Distributor, Generator) individual target functions and a structure formed accordingly.

Actor interactions describe all processes that describe the regional electric power market. There are various types of actor interactions: uni- and bidirectional, continuous and discrete (from the point of view of the considered moments of time), as well as stationary and non-stationary. In what follows, we will consider actor interactions that occur exclusively at discrete times. The system uses two complementary metrics: the eventual and realizable potential of the regional electric power market. The following set of conclusions bases on the results of the above reviews:

1. Risk management processes are quite costly, which requires the introduction of adaptive technologies for launching procedures for stabilizing reconfiguration of the region's energy system in accordance with the emerging information situation.
2. The problem of efficient organization of the regional electric power market is a separate scientific task that requires deep fundamental study, including in the direction of finding effective management of the balance of basic actor interactions.
3. For adequately visualization of current situation, it is necessary to propose effective approaches that allow the decision-maker to provide an aggregated assessment of the current state of the energy system, which is a prerequisite for automated operational decision-making on managing the region's renewable electricity market in real time.

The results of the analysis of existing approaches in the direction of identification, assessment, visualization and management of the regional electric power market allow us to draw a conclusion about the relevance of developing approaches that take into account the features of proactive and reactive control while ensuring actor equilibrium. The next section provides a model for the formal description of the dynamics of the life cycle of a cloud environment under proactive and reactive control.

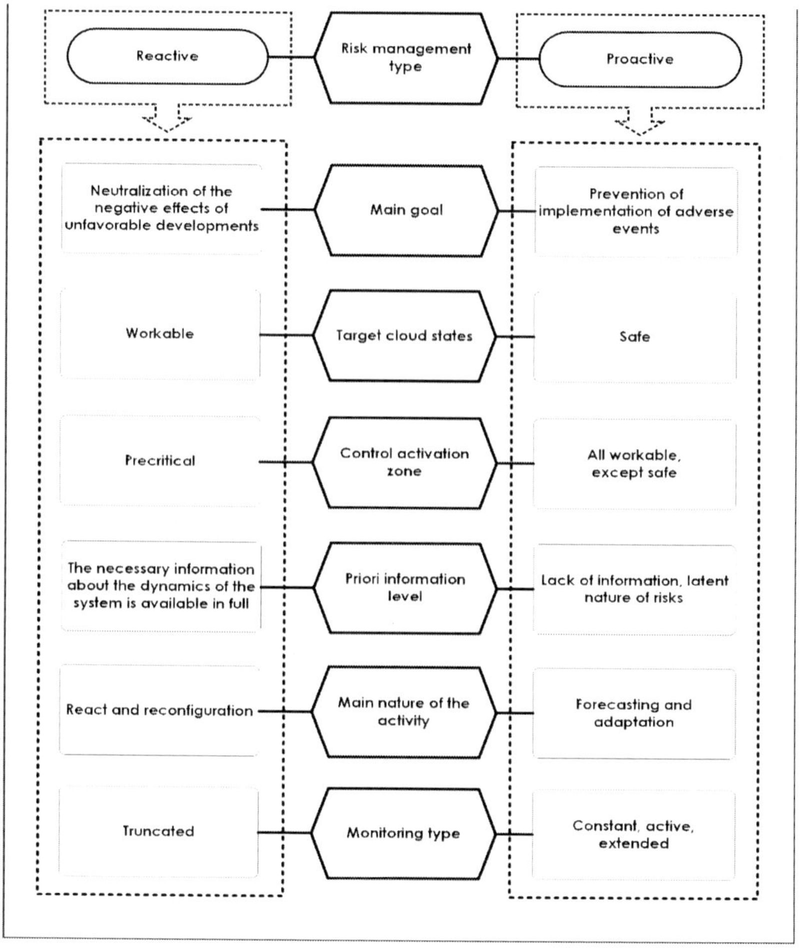

Figure 1. Key features of proactive and reactive energy management in the region.

METHOD

Proposed Methodology for Evaluating the Functional Safety Analysis

The main goal of renewable energy is to ensure a high level of quality and reliability of energy supply to consumers without causing environmental damage to the environment. At the same time, a change in the balance of the electricity market towards green energy can lead to negative consequences - an increase in tariffs and a decrease in the energy intensity of the region. Each energy complex has a hierarchical structure and consists of separate services. The matrix (1) describes service identifiers.

$$\text{USL}=\begin{bmatrix} usl_{11} & \dots & usl_{1m} \\ \dots & \dots & \dots \\ usl_{n1} & \dots & usl_{nm} \end{bmatrix}, \tag{1}$$

where usl_{ij} - identifier of the j-th service of the i-th service; n – number of services in the energy complex of the region; m – number of services in the service.

The states of the energy complex are described using the set S, moreover: $S = S^{safe} \cup S^{oper} \cup S^{prc} \cup S^{abs}$, $S^{safe} \cap S^{oper} \cap S^{prc} \cap S^{abs} = \emptyset$, subsets $S^{safe} = \left\{S_1, S_2, \dots, S_{|S^{safe}|}\right\}$, $S^{oper} = \left\{S_i, \dots, S_k, \dots, S_{|S^{oper}|}\right\}$, $S^{prc} = \left\{S_i^*, \dots, S_k^*, \dots, S_{|S^{prc}|}^*\right\}$, where S^{safe} – is a subset of safe states of the energy complex; S^{oper} – is a subset of operational states of the energy complex; S^{prc} – is a subset of pre-critical states of the energy complex; S^{abs} – is an emergency absorbing condition. From the point of view of the functioning of the energy complex of the region, the transitions between its states can be both constructive and destructive.

The goal of Proactive Management is the following. For $\forall t = \{1, 2, \cdots, w\}$ find such $ur^{opt}(t) = arg \max_{ur_i \in UR^D(t)} \Phi(ur_i(t), t)$, when limiting $\text{S}(ur_i(t), t) \in S^{safe}$, where $UR^{adm}(t)$ – the set of admissible controls at time t; w – number of times in synchronous power distribution control; $\Phi(ur_i(t), t)$ – function evaluating management efficiency $ur_i(t)$ at time t.

The goal of reactive control: $\forall t = \{1, 2, \cdots, w\}$ need to find management $ur^{opt}(t) = arg \max_{ur_i \in UR^D(t)} \Phi(ur_i(t), t)$, while fulfilling the constraints $\text{S}(ur_i(t), t) \notin S^{prc}$, $\text{S}(ur_i(t), t) \notin \notin S^{abs}$.

A general strategic directive take into account the described differences between reactive and proactive approaches. The reconfiguration of the electric power market include the search for such a sequence of controls that would provide the most efficient trajectory of movement from the current state to the desired one, subject to strict observance of the above restrictions. By the trajectory, we mean the sequence of states of the electric power complex of the region when moving from the state S_i to state S_j. The formulations of this problem can vary depending on the availability of a priori information, the approach to control (proactive or reactive) and the chosen solution method. All this requires the creation of a single universal tool that allows making decisions in both automatic and automated modes, effectively using the non-formalized experience of electric power companies.

The organization of risk management in the electric power complex of the region is an essentially non-trivial, contradictory scientific and practical task. The abundance of different approaches to threat analysis, risk assessment and the formation of the necessary countermeasures significantly complicates the already time-consuming task of finding a compromise in the allocation of resources between service and operational processes in critical computing systems. The proposed model of the dynamics of the states of the life cycle of the regional electric power market, in contrast to the existing approaches to formalizing this task, provides an opportunity to integrate various proactive and reactive methods of risk management. It allows the Distributor: on the one hand, to predict the occurrence of risk events, and on the other, to accept operational solutions to minimize losses associated with the occurrence of adverse events. To increase the consistency and conciseness of the presentation of further material, it is advisable to describe the classification notation of the proposed models for ensuring actor equilibrium.

RESULTS AND DISCUSSION

The purpose of the model is to form an equilibrium in the electricity market, which consists in the optimal ratio between the generation of electricity by conventional and renewable energy installations. The model under consideration uses a sufficiently high level of abstraction, which, on the one hand, expands the scope of its application, and on the other hand, restricts the use of the results obtained without preliminary functional and conceptual transformations.

The initial data for it are the parameters characterizing the interactions of the Actors within the framework of the considered service of the services of the electric power complex. The result of its functioning is the obtained balance of the control. Functional blocks are the following:

Φ_B – generates prices for the purchase and sale of electricity by the Distributor;

Φ_C – assesses the amount of electricity that the Consumer is ready to buy at the emerging price;

Φ_P – assess the volume of electrical energy that the Distributor is willing to sell at an emerging price;

Φ_R^c – generates the control actions of the Dispatcher to correct the Distributor's "base" price, determined by the share of renewable energy and regulatory requirements for the power system;

Φ_R^z – generates control actions to correct the Distributor's "markup".

The formally proposed model implements the mapping $\mathfrak{J}^{UA} \to u^{bal}$, where

$\mathfrak{J}^{UA} = \langle c^0, z^0, \Phi_B, \Phi_C, \Phi_P, \Phi_R^c, \Phi_R^z \rangle$ – cortege describing the main parameters of the electricity market;

c^0 – initial price, formed without taking into account the interests of the Distributor;

z^0– the initial value of the Distributor's "margin";

$u^{bal} = \langle c_P^{bal}, c_C^{bal} \rangle$ – a tuple describing the found equilibrium control for the current moment in time;

c_P^{bal}– price that ensures the balance between the Distributor and the Generator;

c_C^{bal}– the price that ensures the balance between the Distributor and the Consumer.

The system (2) formalize functional requirements. Figure 2 shows the organizational and functional structure of the management model of actor balance when providing a cloud service. It is two-channel: the direct channel is implemented by functional blocks Φ_B, Φ_C, Φ_P; feedback - in blocks Φ_R^c and Φ_R^z. The proposed formalization model use the system-dynamic library of the AnyLogic simulation environment. The system include various procedures for finding equilibrium control (PID controller or Alfa controller). The dispatcher set initial conditions $\langle c^0, z^0 \rangle$.

$$\begin{cases} \langle c_P(t), c_C(t) \rangle = \Phi_B\left(\frac{dc}{dt}, \frac{dz}{dt}\right); \\ \frac{dc}{dt} = \Phi_R^c(c^0, W(V_C, V_P, t), \mathrm{u}_R^c); \\ \frac{dz}{dt} = \Phi_R^z(z^0, W(V_C, V_P, t), \mathrm{u}_R^z); \\ V_C(t) = \Phi_C\left(c_C, \frac{dc}{dt}, \frac{dz}{dt}\right); \\ V_P(t) = \Phi_P\left(c_P, \frac{dc}{dt}, \frac{dz}{dt}\right); \\ \mathrm{u}_R^c = \lambda W(V_C, V_P, t) + \beta \int_0^T W(V_C, V_P, t)dt + \gamma \frac{dW(V_C, V_P, t)}{dt}; \\ \mathrm{u}_R^z = \begin{cases} -\alpha W(V_C, V_P, t), \ at \ W(V_C, V_P, t) \geq W_{\mathrm{thr}} \\ \alpha(V_C, V_P, t), \ at \ W(V_C, V_P, t) < W_{thr} \end{cases}; \\ under \ initial \ conditions \ \langle c^0, z^0 \rangle. \end{cases} \quad (2)$$

where c_P – the price formed by the Distributor for the purchase of electricity from the Generator;

c_C – the price formed by the Distributor for the sale of the volume of electricity to the Consumer;

$\frac{dc}{dt}$ – the derivative, on the basis of which the price correction is performed per unit of electricity volume, formed without taking into account the interests of the Distributor in accordance with the corrective control actions received from the Dispatcher;

$\frac{dz}{dt}$ – derivative, on the basis of which the Distributor's "margin" is formed per unit of the supplied volume of electricity.

W – electricity sales (system characteristic);

V_C – the amount of electricity that the Consumer is willing to purchase at a price c_C;

V_P – the amount of electricity that the Generator is ready to sell at a price c_P;

u_R^c – control actions of the price correction Dispatcher c;

u_R^z – Distributor / Dispatcher control actions for margin correction z;

λ, β, γ – the gains of the proportional, integrating and differentiating components of the PID controller, which can be found using adaptive selection procedures or the use of neural networks;

α – Alfa controller coefficient;

W_{thr} – sales threshold W for Alfa controller.

The dynamics of V_C and V_P changes for the developed model is described as follows:

$$V_C^{i,j}(t, \Delta t) = V_C^{i,j}(t_0, \Delta t) \cdot \alpha^{i,j}(t) + c_C^{i,j}(t_0, \Delta t) \cdot \beta^{i,j}(t);$$
$$V_P^{i,j}(t, \Delta t) = V_P^{i,j}(t_0, \Delta t) \cdot \gamma^{i,j}(t) + c_P^{i,j}(t_0, \Delta t) \cdot \mu^{i,j}(t),$$

where $\alpha^{i,j}(t), \gamma^{i,j}(t)$ – coefficients that take into account the "inertia" of the Actors when considering the j-th service of the i-th service of the electric power complex;

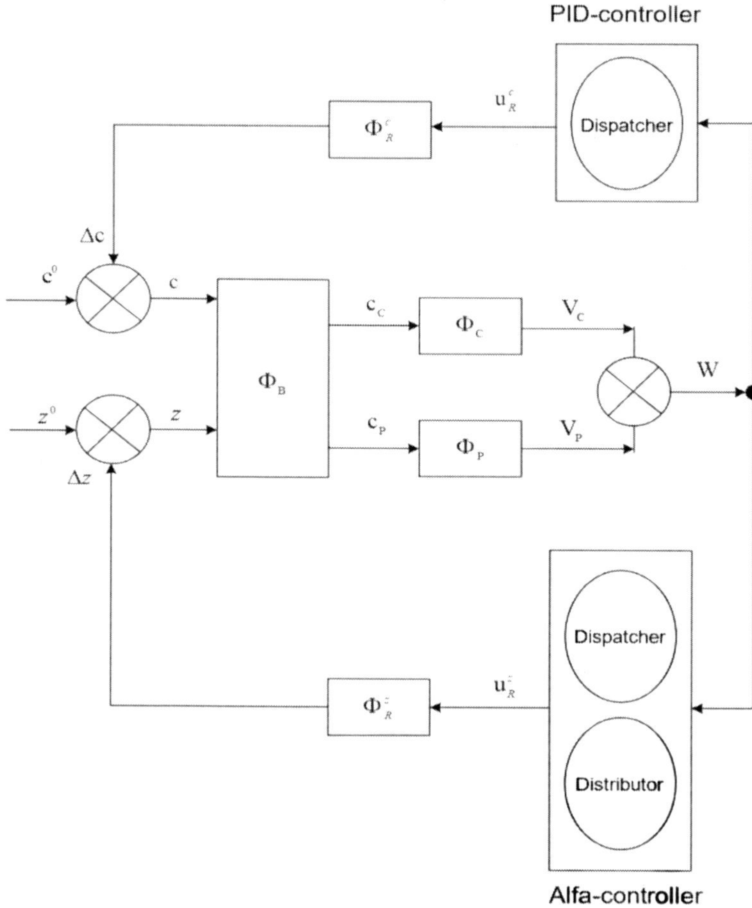

Figure 2. Organizational and functional structure of the model for managing the actor equilibrium of the electricity market.

$\beta^{i,j}(t)$, $\mu^{i,j}(t)$ – coefficients that take into account the "elasticity" of the Actors when considering the j-th service of the i-th service of the electric power complex.

It is obvious that:

$$\frac{\Delta V_C^{i,j}(t)}{\Delta t} = V_C^{i,j}(t_0) \cdot \alpha^{i,j}(t) + c_C^{i,j}(t) \cdot \beta^{i,j}(t);$$

$$\frac{\Delta V_P^{i,j}(t)}{\Delta t} = V_P^{i,j}(t_0) \cdot \gamma^{i,j}(t) + c_P^{i,j}(t) \cdot \mu^{i,j}(t).$$

Thus, the system of differential equations (3) with the initial conditions (4) describes the dynamics of the Consumer's demand. The system of differential equations (5) with the initial conditions (6) give the proposal for the Generator.

The following are dialog boxes for software implementation of the model when using the AnyLogic simulation environment. It is possible to set parameters that determine Φ_C, Φ_B, Φ_P, Φ_R^C, Φ_R^Z. The implemented alternative procedures for finding equilibrium control have been are the following: PID controller, Alfa controller. Figure 3 shows a system dynamic model of the regional electric power market, developed in the AnyLogic program, and Figure 4 shows graphs of the results of its work.

$$\begin{cases} \frac{dV_C^{1,1}(t)}{dt} = \Phi_C(1,1,\alpha^{1,1}(t),\beta^{1,1}(t),t); \\ \dots \\ \frac{dV_C^{i,j}(t)}{dt} = \Phi_C(i,j,\alpha^{i,j}(t),\beta^{i,j}(t),t); \\ \dots \\ \frac{dV_C^{n,m}(t)}{dt} = \Phi_C(n,m,\alpha^{n,m}(t),\beta^{n,m}(t),t). \end{cases} \tag{3}$$

$$\begin{cases} V_C^{1,1}(t_0) = \eta_C^{1,1}; \\ \dots \\ V_C^{i,j}(t_0) = \eta_C^{i,j}; \\ \dots \\ V_C^{n,m}(t_0) = \eta_C^{n,m}. \end{cases} \tag{4}$$

$$\begin{cases} \frac{dV_P^{1,1}(t)}{dt} = \Phi_P(1,1,\gamma^{1,1}(t),\mu^{1,1}(t),t); \\ \dots \\ \frac{dV_P^{i,j}(t)}{dt} = \Phi_P(i,j,\gamma^{i,j}(t),\mu^{i,j}(t),t); \\ \dots \\ \frac{dV_P^{n,m}(t)}{dt} = \Phi_P(n,m,\gamma^{n,m}(t),\mu^{n,m}(t),t). \end{cases} \tag{5}$$

$$\begin{cases} V_P^{1,1}(t_0) = \eta_P^{1,1}; \\ \quad\quad \dots \\ V_P^{i,j}(t_0) = \eta_P^{i,j}; \\ \quad\quad \dots \\ V_P^{n,m}(t_0) = \eta_P^{n,m}. \end{cases} \tag{6}$$

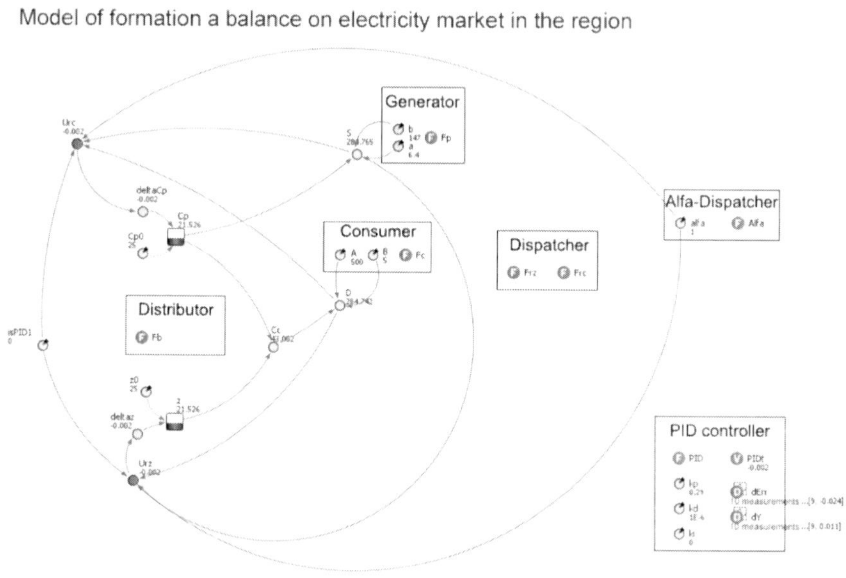

Figure 3. System-dynamic model of the regional electricity market.

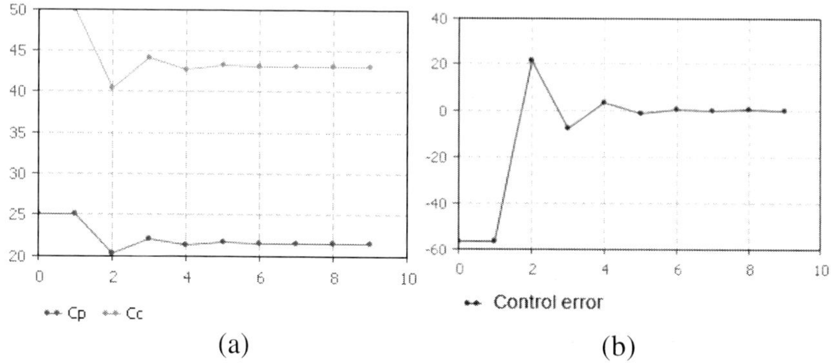

Figure 4. Graphic dependencies: (a) dynamics of change in Cp (blue line) / Cc (red line) depending on the model time; (b) dynamics of change of control error depending on model time

After a certain number of conventional units of the model time, the formed self-oscillations decay and the necessary control ensure equilibrium. The proposed approach, in contrast to the known ones, is implemented using the system-dynamic library of the AnyLogic simulation environment, the PID controller and is focused on using within the extended reference model of the regional electric power market. This scientific result is interdisciplinary, allows one to abstract from a specific engineering implementation (which expands the scope of its application), but at the same time describes the process of finding a compromise in actor interaction at a sufficient level of detail acceptable for making systemic decisions to overcome collisions arising in the organization of the power grid.

The results obtained using this model are a special case of equilibrium control and are valid in the framework of only a single situation. However, the main goal is to develop a compromise option for organizing the entire electric power complex, balanced both taking into account the interests of the basic Actors within the framework of individual services of each service, and taking into account resource collisions between various services of the electric power market.

CONCLUSION

The main result of the study is the improvement of the systematic approach for calculating the optimal parameters of the energy complex of the region, based on the use of traditional and renewable energy resources. The choice of a strategy for the development of the energy market in the region is a substantially non-trivial task. It requires serious information support both from the point of view of interactive decision support technologies and the use of powerful predictive analytics procedures based on the use of hybrid simulation, advanced dual control methods, and agent-based simulation when looking for compromise solutions that balance the interests of various actors in the electricity market in the region. The proposed approach bases on the use of a modified regulator, which, unlike known solutions, allows taking into account the contradictions between actors not only at the current moment of time, but using the prehistory of the behaviour of actors in the market. It also analyses the dynamics of the growth of contradictions in order to use a proactive approach to management in advance and minimize the number of conflicts between actors. The results obtained demonstrate the effectiveness of the proposed approach, as well as form the directions for its further development. It is useful to deepen the use of the proposed mathematical apparatus from the point of view of taking into account the specific details of the subject area under consideration. In addition, it is planned to increase the efficiency of the decisions made, as well as the depth of the forecast with specialized high-performance computing facilities.

The proposed and implemented information technology for decision-making support for organizing actor interactions in the regional electric power market uses the AnyLogic multi-approach simulation environment. It focuses on the organization of computational services for large-scale scientific problems. It also integrates the methods of network analysis of projects, agent-based and discrete-event paradigm for the synthesis of simulation models of complex systems, which allows decision makers to increase the efficiency of decisions made in the framework of the extended reference model of cloud computing. The results of numerical modelling carried out that the using of the proposed approach for most regions is incorrect. Share of power generating plants using renewable energy resources should be about 25% for regions with a share of industrial consumers constituting less than 40%, and for other regions no more than 16%.

The abundance of different approaches to threat analysis, risk assessment and the formation of the necessary countermeasures significantly complicates the already time-consuming task of finding a compromise in the allocation of resources between service and operational processes in critical computing systems. The proposed formalization of this task makes it possible to integrate various proactive and reactive methods of risk management of cloud systems, which allows the Broker: on the one hand, to predict the onset of risk events, and on the other hand, to make operational decisions to minimize losses associated with the occurrence of adverse events. In particular, the proposed information portrait allows visualizing the aggregated risk assessment for the current state of the cloud-computing environment, which is a prerequisite for making effective decisions in real time. The developed complex of models for effective management of the equilibrium of actor interactions in Cloud systems uses a dual approach and uses the consideration of economic processes in cybernetic systems.

Future research directions of the proposed approach contain development and implementation of methods for forecasting and dispatching energy consumption, contributing to the sustainable development of the regional energy system, including a self-learning element in the digital model to enable automatic analysis of emerging information situations in the regional energy market. The practical use of this technology reduces the risk of a monopoly on the energy market in the region.

ACKNOWLEDGMENTS

The study was carried out with partial financial support from the Russian Foundation for Basic Research and the Government of Sevastopol in the framework of the scientific project № 20-47-920006, as well as the Sevastopol State University in the framework of the internal grant № 35/06-31 under the project "Application of big data technologies for solving problems of environmental monitoring" and № 27/06-31 under the project "Development of competitive technologies for creating components of a mobile photovoltaic installation of high efficiency".

REFERENCES

Ahmad, T., Zhang, H., & Yan, B. (2020). A review on renewable energy and electricity requirement forecasting models for smart grid and buildings. *Sustainable Cities and Society*, 55, 102052.

Alhendawi, K. M., Al-Janabi, A. A., & Badwan, J. (2019, October). Predicting the Quality of MIS Characteristics and End-Users' Perceptions Using Artificial Intelligence Tools: Expert Systems and Neural Network. In *International Conference on Intelligent Computing & Optimization* (pp. 18-30). Springer, Cham.

Belyaev, L.S., Marchenko, O.V., & Podkovalnikov, S.V. (2002). Power Industry in Russia: Consequences of the Transition to a Competitive Market. *Energy: economics, technology, ecology*, (6), 2-9.

Bogachkova, L. Yu. (2018). Increasing energy efficiency as a driver of the global competitiveness of the national economy. Economics and Management: *Theory and Practice*, 4 (1), 22-31.

Borenstein, S., Bushnell, J., & Knittel, C. R. (1999). Market power in electricity markets: Beyond concentration measures. *The Energy Journal*, 20(4).

Clayton, T., & Radcliffe, N. (2018). *Sustainability: a systems approach.* Routledge.

Dittmar, M. (2012). Nuclear energy: Status and future limitations. *Energy,* 37(1), 35-40.

Elluru, S., Gupta, H., Kaur, H., & Singh, S. P. (2019). Proactive and reactive models for disaster resilient supply chain. *Annals of Operations Research*, 283(1), 199-224.

Fu, T. Z., Ding, J., Ma, R. T., Winslett, M., Yang, Y., & Zhang, Z. (2015, June). DRS: Dynamic resource scheduling for real-time analytics over fast streams. In *2015 IEEE 35th International Conference on Distributed Computing Systems* (pp. 411-420). IEEE.

Geraskin, M. (2017). Game-theoretic analysis of Stackelberg oligopoly with arbitrary rank reflexive behavior of agents. *Kybernetes.*

Ines, C., Guilherme, P. L., Esther, M. G., Swantje, G., Stephen, H., & Lars, H. (2020). Regulatory challenges and opportunities for collective renewable energy prosumers in the EU. *Energy Policy*, 138

Kapitonov, I. A. (2019). Energy security of territories as a factor of sustainable development under the conditions of economic changes, *International Journal of Energy Economics and Policy 9(6), 210-221*

Kuznetsov, P. N., Lyamina, N. V., & Yuferev, L. Y. (2019). A device for remote monitoring of solar power plant parameters. *Applied solar energy*, 55(4), 247-251.

Leuthold, F. U., Weigt, H., & von Hirschhausen, C. (2012). A large-scale spatial optimization model of the European electricity market. *Networks and spatial economics*, 12(1), 75-107.

Li, X., Salehi, M. A., Joshi, Y., Darwich, M. K., Landreneau, B., & Bayoumi, M. (2018). Performance analysis and modeling of video transcoding using heterogeneous cloud services. *IEEE Transactions on Parallel and Distributed Systems,* 30(4), 910-922.

Lospinoso, J. A., Schweinberger, M., Snijders, T. A., & Ripley, R. M. (2011). Assessing and accounting for time heterogeneity in stochastic actor oriented models. *Advances in data analysis and classification*, 5(2), 147-176.

Ma, T., Guo, C., Cao, Y., Han, Z., Qin, J., & Zhang, W. (2010). Current status and development trend of intelligent dispatching automation system of power system. *Dianli Xitong Zidonghua (Automation of Electric Power Systems)*, 34(9), 7-11.

Mitiku, T., & Manshahia, M. S. (2020, December). Adaptive Neuro-Fuzzy Inference Based Modeling of Wind Energy Harvesting System for Remote Areas. In *International Conference on Intelligent Computing & Optimization* (pp. 3-18). Springer, Cham.

Monteiro, C., Ramirez-Rosado, I. J., Fernandez-Jimenez, L. A., & Conde, P. (2016). Short-term price forecasting models based on artificial neural networks for intraday sessions in the Iberian electricity market. *Energies*, 9(9), 721.

Rau, N. S. (2003). *Optimization principles: practical applications to the operation and markets of the electric power industry* (Vol. 16). John Wiley & Sons.

Roche, R., Blunier, B., Miraoui, A., Hilaire, V., & Koukam, A. (2010, November). Multi-agent systems for grid energy management: A short review. In *IECON 2010-36th Annual Conference on IEEE Industrial Electronics Society* (pp. 3341-3346). IEEE.

Samakpong, T., Ongsakul, W., & Madhu, M. N. (2019, October). Optimal Power Flow Considering Cost of Wind and Solar Power Uncertainty Using Particle Swarm Optimization. In *International Conference on Intelligent Computing & Optimization* (pp. 190-203). Springer, Cham.

Tan, C., & He, J. (2021). Integrated proactive and reactive strategies for sustainable berth allocation and quay crane assignment under uncertainty. *Annals of Operations Research*, 1-32.

Tao, Q., Liu, F., & Sidorov, D. (2019, October). Recurrent Neural Networks Application to Forecasting with Two Cases: Load and Pollution. In *International Conference on Intelligent Computing & Optimization* (pp. 369-378). Springer, Cham.

Vasil'ev, V. A. (2016). A-equilibrium and fuzzy A-core in pure exchange model with externalities. *Automation and Remote Control,* 77(11), 2080-2089.

Voropai, N.I., Trufanov, V.V., & Sheveleva, G.I. (2004). Ensuring investment and development of the Russian electric power industry during the transition to a competitive market. *Energy Policy*, (2), 44.

Wadhwa, C. L. (1989). Generation, distribution and utilization of electrical energy. *New Age International.*

Yaghooti, A., Buygi, M. O., & Shanechi, M. H. M. (2015). Designing coordinated power system stabilizers: a reference model based controller design. *IEEE Transactions on Power Systems*, 31(4), 2914-2924.

Yuferev, L., & Kuznetsov, P. N. (2019). Increasing the Energy Efficiency of Photovoltaic Systems Operating Under Conditions of Uneven Illumination. In *Renewable Energy and Power Supply Challenges for Rural Regions*, 88-114

Yuvaraj, D., Sivaram, M., Ahamed, A. M. U., & Nageswari, S. (2019, October). An efficient lion optimization based cluster formation and energy management in WSN based IoT. In *International Conference on Intelligent Computing & Optimization* (pp. 591-607). Springer, Cham.

In: Advances of Machine Learning … ISBN: 978-1-68507-211-7
Editors: Pandian Vasant et al.

Chapter 6

RES-BASED POWER PLANTS VERSUS POLLUTING POWER PLANTS: PROS AND CONS

Vytautas Adomavicius*
Department of Electric Power Systems,
Kaunas University of Technology, Kaunas, Lithuania

ABSTRACT

This paper presents a review of the presently used main electric power generation technologies: three of them are based on renewable energy sources (solar, wind, and hydropower plants) and the remaining three on the fossil fuels (natural gas, coal, and nuclear power plants). The most important factors of all power plants are presented for the comparison of their properties. The disadvantages and advantages of all types of power plants are described. The information provided in the paper helps us to understand why the RES-based power plants currently are experiencing breakthroughs in the top positions in terms of electricity generation worldwide and in terms of the lowest indicators of environmental pollution. The information found in the reviewed references and provided in this paper along with the author's insights lead

---

* Corresponding Author's Email: vytautas.adomavicius@ktu.lt.

to the conclusion that this breakthrough in RES-based technologies should be successful and that polluting power generation technologies will gradually be abandoned due to their significantly lower economic and environmental performance. There is a good chance that the green revolution in the global energy sector will succeed.

Keywords: RES-based power plants, polluting power plants, key indicators, efficiency of operation, comparison of the technologies, prospects for future development

List of Notations and Abbreviations

APVS	Agro-photovoltaic system
CO_2	Carbon dioxide
CPP	Coal power plant
GW	Gigawatt
HEPP	Hydroelectric power plant
kW	Kilowatt
kWh	Kilowatt-hour
LCOE	Levelized cost of electricity
LCPP	Low-carbon power plants
MW	Megawatt
MWe	Megawatt of electric power
MWh	Megawatt hour
NGPP	Natural gas power plant
NPP	Nuclear power plant
OPP	Oil-based power plants
ORPP	Other renewable power plants
PPP	Polluting power plants
PV	Photovoltaic
PVPP	Photovoltaic power plant
RES	Renewable energy sources
RESPP	RES-based power plants
SMR	Small modular reactor

TGC	Total global (operating) capacity
TW	Terawatt
TWh	Terawatt-hour
WF	Wind farm
WHO	World Health Organization
WPP	Wind power plant.

INTRODUCTION

Various technologies of electricity and thermal energy production are constantly competing with each other. Especially interesting is the second decade of 21st century when the competition escalated dramatically the world over. A very clear and powerful breakthrough can presently be seen in those energy production technologies that use renewable energy as their primary source. First of all, the winners in this competition are determined by the already better and constantly improving economic indicators of the RES-based energy production technologies. Moreover, the incomparably better ecological performance of these technologies is of paramount importance, because most people no longer want to breathe air, 'flavoured' with harmful pollutants from power plants, industry, internal combustion vehicles, and agricultural machinery powered by fossil fuels. These people do not want themselves, their children, grandchildren, and the next generations to be victims of climate change in many aspects.

Many people already understand the need to preserve clean nature, to clean up the air that everyone has to breathe, protect health from contaminants in order to avoid high healthcare costs due to the adverse effects of toxic pollutants, and to stop climate change, which is already causing huge material damage. However, many uninformed people do not rely on scientists. People interested in supporting environment-polluting, power generation technologies are trying to prolong the existence of already obsolete technologies in various ways for their personal reasons. Of course, all this may slow down the process of the green energy revolution a little, but they will not achieve their goals, because, in free democracies, new, more advanced, and in all respects more efficient technologies always win according to economic laws. Other options are only available in some countries led by 'all-knowing' dictators, where scientific advice is quite often ignored. Opponents of modern energy technologies usually have sufficient funds to introduce innovations, so they could hopefully be advised to initiate a gradual shift towards new, more advanced green power generation technologies. Some well-known companies are already going in this direction.

In a democratic society, everyone can have their own opinion on any issue and it does not matter whether his or her opinion is scientifically sound due to lack of education, or whether it is deliberately misleading for selfish ends. The most important circumstance for proper decision-making is when the science-based decision is understood and supported by the vast majority of citizens. Often, very important decisions have to be made quickly, so there is no time to wait until everyone understands all sides of the issue and hopefully everyone will support the decision in good faith. There will always be different opinions, but this cannot prevent decision-making – much like trains are not waiting for those lagging behind.

This chapter presents the top three most promising RES-based power plants (RESPP), which are compared to the corresponding indicators of the three most important power plants running on fossil fuels. A review of the advantages and disadvantages of all types of power plants has been carried out and their prospects for further development in the near future have been examined in order to find the best way to go. The suitability of the aforementioned power generation technologies for rural development has also been taken into consideration.

RES-Based Power Plants

The main RES-based electricity generating sources are the hydroelectric power plants, wind power plants, and solar power plants (both photovoltaic and solar thermal). There are some other renewable energy plants having a moderate impact on global electricity production, being geothermal, biomass, waste, wave, and tidal power plants. Shares of the RESPP in the total global power production in the year 2020 were as follows [1]:

- Hydroelectric power plants (HEPP) – 16.85%;
- Wind power plants (WPP) – 6.15%;
- Solar power plants (SPP) – 3.27%;
- Other renewable power plants (ORPP) – 2.72%.

Solar, wind, and hydropower plants were selected for further examination and comparison of their most important indicators with those of the three types of polluting power plants. HEPP already accounts for the significant share of global electricity, while solar and wind power plants currently show unprecedented expansion rates, the lowest levelized costs of electricity, and the lowest indicators of environmental pollution.

RESPP of all types and nuclear power plants (NPP) make up a group of so-called low-carbon power plants (LCPP). Simultaneously, the NPPs can be considered as polluting power plants. Pollution of the environment by the NPPs can be considered bearable if the power plant is operating normally. However, in the event of a major accident, environmental pollution and material damage can be enormous.

Solar Power Plants

According to the World Energy Council's estimation, solar radiation has the greatest potential for electricity generation – minimal global usable solar energy potential is 444,800 TWh, maximal is 13,884,4000 TWh [2]. At the beginning of the solar power plant era, while PV modules were expensive, the world was dominated by the viewpoint that PVPP as a significant power generation technology was not viable. It was true especially in rural areas not only because of the high price of solar power, but also because the majority of decision-makers thought that all land must first be used for producing agricultural products which are chronically lacking in the world nowadays.

However, over time, the efficiency indicators of solar power plants have greatly improved. At present, they are already significantly better than all major fossil fuel power plants and are improving further. The levelized cost of electricity (LCOE) of the photovoltaic power plants (PVPP) according to Lazard data is currently the lowest among all power generation technologies. Onshore wind power plants (WPP) and hydroelectric power plants (HEPP) are slightly behind the PVPP according to the LCOE (PVPP – 37 USD/MWh, WPP – 40 USD/MWh, and HEPP – 50 USD/MWh) [3].

A very large advantage of solar power plants is the fact that these power plants do not require any fuel for electricity production. Unlike power plants running on fossil fuels, there is no need to worry about fuel supply. There should be no political pressure on the supply of various types of energy resources or their cost, because the energy, in the form of sunlight, arrives at the installation sites of these power plants in a very short time after the moment when the sunlight is generated. The speed of light in a vacuum (in space) is defined as 299,792,458 m/s. The distance from the Sun to the Earth varies between 146,000,000 and 152,000,000 km. Consequently, the light born in the Sun arrives on Earth between 8 min 7 s and 8 min 27 s (the average time is 8 min 17 s) depending on the season of the year. Solar rays have always been and will continue to be free.

Therefore, PVPP is one of the most environmentally friendly ways to produce electricity. The environmental impact is minimal as they do not emit harmful pollutants, including greenhouse gases. They are not generating noise because they do not have moving parts. They do not use water for electricity production. The water in many countries is already becoming deficient (especially drinking water). Small amounts of water in these power plants can be used to wash pollinated solar modules if they are not washed away by rain for a long time.

Solar power plants are very simple, quick, and easy to install and dismantle compared to traditional polluting power plants. They do not require very highly qualified specialists. Small-scale solar power plants can be installed and maintained by almost any trained electrician. The power range of solar power plants is very wide – from milliwatts (mW) to gigawatts (GW). The government of Australia's Northern Territory and Sun Cable Company of Singapore have signed a deal in January 2021 to develop the Australia-ASEAN Power Link. A huge 10 GW solar power farm will be built in the Northern Territory and electricity from this farm will be supplied via high voltage direct current submarine cable to Singapore [4]. Solar power plants had the highest added capacity in 2020 among all power generation technologies used in the world – 127 GW [5].

Solar power plants of suitable power can be designed and installed in many places in almost every area and almost every house. Solar modules are often installed where they do not need additional space: on the roofs of buildings, on walls, above parking lots (like a canopy), and on bodies of water. Solar modules integrated into the pavement of motorways, bicycle roads and elsewhere have already been tested. Solar modules are also mounted on the top of some electric vehicles, on electric boats, catamarans, ferries, drones, and small planes, so that their batteries need to be recharged less frequently. Sufficiently transparent photovoltaic modules that can be used in the windows of large buildings instead of glass and photovoltaic modules-tiles for roofing have also been developed. The price of the tiles is not significantly different from that of ordinary tiles that do not produce electricity. Small modules are also fastened on some outerwear, backpacks, appliances, tents, and elsewhere. The saying "put solar on it!" has spread around the world.

Solar power plants are very suitable for installation in rural areas, especially when the LCOE of the PVPP became the lowest among all power production technologies and when the concept of double land use is applied. This concept was proposed by German scientists from the Fraunhofer Institute, A. Goetzberger and A. Zastrow in 1981. They proposed to use the land for two purposes simultaneously – for agriculture and for electricity generation by means of PVPPs and therefore this combination was named the agro-photovoltaic system (APVS, presently agri-voltaic and agrivoltaic is also used). This concept was described for the first time in their paper [6] published in 1982. Unfortunately, there were no favourable conditions for the implementation of this idea at the time as solar modules were still very expensive. Favourable conditions for the implementation of this idea appeared only in the second decade of the current century.

APVSs are also worthwhile to be installed on fertile lands used for agricultural purposes. Scientists from the Fraunhofer Institute in Germany carried out a study in 2018 in order to find out the effectiveness of the exploitation of a plot of land when the APVS is installed on that site [7]. Agro-photovoltaic systems were tested in the experimental plots of land (1 ha), where various agricultural crops were grown and electricity was produced using solar modules. There were also control plots of land (1 ha) nearby, but there was no double use of the land. It was for agricultural crops only. Solar modules were installed on one such experimental plot and potatoes were planted there too. The same type of potato was planted on a control soil plot nearby. After harvesting the potatoes on both plots, it was found that in the agro-photovoltaic system, the yield of potatoes was only 14% lower than on the reference plot, and the potatoes planted had no impact on electricity production at this solar power plant. In this way, 186% land use efficiency was achieved on this experimental plot.

Solar modules can also be installed on less fertile land, as well as in wastelands and deserts that are not used for agricultural production. If arrays of solar modules are fenced, it is purposeful to keep sheep, goats, rabbits, bees, chicks, geese, or other suitable animals inside those enclosures. There is no need to mow the grass when sheep, goats, or deer graze on the plots of the solar arrays, as these animals eat it. Lamb rearing in a solar pasture (containing PVPP) has been extensively studied by researchers at Oregon State University in the US in 2019-2020 [8]. The results of the research show that the land equivalent ratio (LER) for the live weight production was 1.81 (181%) in 2019 and 2.04 (204%) in 2020. It can be assumed that lambs liked spending the time ruminating and idling in the shade under the PV array - they spent 96% of the time in these activities. 45% of grazing time was spent under the array too where the value of grass was found to be better as it was less dry. Economic analysis of this APVS is presented in the paper [9].

In addition, solar power plants have much more useful properties. Their owners can install a power plant of practically any capacity, as long as they have sufficient funds and space (on the ground, on the roof, in the water body, on the vehicle, and elsewhere). Solar arrays can be mounted on various water bodies, such as reservoirs of hydroelectric power plants (HEPP), lakes, ponds, and even on the surface of the seas in places not reached by strong winds. Solar arrays do not occupy the ground in these cases, reduce water temperature and evaporation rate in HEPP ponds, the temperature of solar modules, so both types of power plants (hydro and solar) can produce more electricity.

Distributed electricity generation also has its advantages. In the cases, when PVPP is installed on the site for self-consumption (e.g., building, office, company, farm, etc.) and the amount of energy produced approximately meets the needs of all local power receivers, most of which are consumed during the day, significant power losses because of the power transmission and distribution can be avoided. It is preferable to use more reliable microinverters [10] for this type of solar power plant instead of the string inverters. Nowadays, every electricity consumer can become a prosumer by installing a power plant that meets its energy needs and pays off after a few years.

Small-scale PVPPs are best suited to implement the idea of distributed electricity generation and to reap the benefits of this technology. This leads to a significant reduction in environmental pollution and significant electricity losses in traditional old types of electricity systems during the generation, transmission and distribution phases. Compared to large fossil fuel power plants, they also have more advantages, among which the following are important:

- The idea of full or partial energy independence becomes realistic for individual electricity consumers;
- The security of the energy supply is significantly enhanced nationwide in the event of natural disasters, terrorist attacks or

war, if large-scale distributed electricity generation is implemented;
- New small-scale PVPP is installed much faster and cheaper. The energy supply can be restored if it breaks down for various reasons, including natural disasters, and old plants can be dismantled much faster.

One of the most significant drawbacks of PVPP is the frequent variation of solar irradiance depending on the time of day and the season. As a result, they do not produce electricity at night. Luckily, the electricity needs then are low. During the winter period, they produce much less electricity than during the summer period. Supplying significant unstable power into the grid of the power system may make power balancing a little more difficult for the operator of the power grid when the total capacity of all PVPP integrated into the system is large enough.

Wind Power Plants

Wind power plants (WPP) and PVPP are currently the dominant power generation technologies among all newly built power plants. Last time their total capacity installed per year accounted for about 80% of the capacity of all new power plants built during the same year and the WPP was the leading technology in this area. WPP had the second-highest added capacity in 2020 among all power generation technologies – 111 GW [11] after solar power plants. The total global operating capacity of WPP was 0.734 TW in the year 2020 and it was the fourth power-generating technology in the world after the CPP, NGPP, and HEPP. The pace of development of solar and wind power plants is now so rapid that, according to forecasts, they will begin to dominate the electricity generation market as early as this decade [12]. Other key WPP indicators are also good and similar to PVPP indicators: LCOE of wind power plants is slightly higher ($40/MWh) [3], carbon intensity is slightly lower (12 g CO_2 eq/kWh) [13] and installation cost is higher ($1,846-4,362/kW) [14] and depends on the WPP type (onshore or offshore).

Power generation technology using WPP also has some undesirable properties. The most important of them is that the wind speed is very erratic, which is the reason why the electricity generated by WPP is constantly changing. This feature slightly complicates the supply of electricity produced by powerful wind farms (WF) to the grid of the electricity system. As in the case of the PVPP, problems of unstable energy integration into the electricity grid are successfully addressed using modern technologies of automatic control, IT, and AI.

Opponents of the WPP often like to stress avian mortality because of the wind turbines. Yes, a certain amount of birds are killed, however, a number of scientific studies carried out in order to have reliable information on this issue showed that WPPs are "lagging behind" in a long list of avian mortality reasons. The study carried out in the US and Europe in the first decade of the 21st century estimated that WF and NPP are almost equally responsible for avian killings. They have estimated about 0.3-0.4 fatalities per GWh [15]. However, PPPs are much more dangerous – they are responsible for about 5.2 fatalities per GWh. According to the study, WF in the US in the first decade killed 7,000 birds, NPP – 327,000, and the PPP based on fossil fuels – 14,500,000 birds. According to another reference, annual avian mortality in the USA is mostly caused by power transmission lines (174,000,000 birds per year), domestic and feral cats (210,000,000), buildings and windows (365,000,000), use of pesticides (72,000,000), agriculture (67,000,000), cars and trucks (50,000,000), power plants running on fossil fuels (14,000,000) and other reasons. WPP ranks only 11th on this list with the number of 140,000 birds per year [16].

Hydroelectric Power Plants

Hydropower plants (HEPP) currently are one of the leading power generation technologies widely used since the 19th century. Its energy and environmental performance is among the best compared to other technologies: LCOE – $50/MWh (third among all technologies) [17]; total global operating capacity – 1,324 TW (third) [18]; added global power per 2020 – + 84 TWh (third) [19]; carbon intensity – 4 g CO2 eq/kWh (the best) [13]. The most powerful electric power plant in the world among all power generation technologies is HEPP Three Gorges built in China (22.5 GW) [20].

HEPP and PPP are friendly to each other. Floating PVPP can do a lot to help the HEPP generate electricity together from the same pond. HEPP built in plains have large areas of the upper pond and the floating PVPP can often produce even more electricity than HEPP itself. Such a hybrid hydro-solar power plant can be used to balance power in the power transmission network. In this case, the power in the electricity system can be balanced by regulating the water level in the upper pond and the number of hydro units in operation.

Hydroelectric power plants also derive some added benefit from this friendship. When large areas of the upper pond are covered by floating solar arrays, less water evaporates from the pond per year and more water flows through the hydro turbines. Also, the water in the pond heats up slightly less and the conditions for algae to reproduce deteriorate.

One of the most important shortcomings of HEPP is the fact that the hydroelectric dam prevents valuable fish species (e.g., salmon) from entering their upstream spawning grounds. Fish ladders are functioning satisfactorily only in dams of small height. In high-altitude dams, the ladders do not create good conditions for fish to overcome this barrier, and their installation is expensive.

The erosion of the banks of their upper reservoirs is another disadvantage of these power plants. Strong winds in reservoirs of a large area make waves that rinse the banks of the reservoir, slowly destroy and wash them into the water. Thus, over the years, tens and even hundreds of meters of land can be washed into the water.

In addition, electricity generation in the HEPP requires the largest share of water losses to produce 1 MWh of electricity between all power generation technologies. Part of the water evaporates from the upper reservoir of the power plant, and the other part of the water is absorbed into the ground at the bottom of the dam. As a result, the power plant produces slightly less electricity, since lost water does not pass through the turbines. The lost share of water depends on the type of dam, on the area occupied by the surface of the water, and on the characteristics of the soil at the bottom of the dam. Large areas of the upper reservoir per megawatthour produced have HEPPs operating in the plains, and as a result, losses of water there are more substantial than in the HEPPs built in mountains. On average, the amount of water lost per MWh produced is about 17 m^3 [21].

Environmentalists often name another drawback of HEPP - every summer, water in the upper reservoir of the power plant "blooms" for a period of 2-4 weeks, when favourable conditions for algae to multiply in the pond are created and large areas of the surface of the water are covered by a stinking layer of green algae. Then the number of holidaymakers and athletes near the pond decreases significantly. Abundantly flowering algae intensively uses oxygen dissolved in water. In case of a lack of oxygen in water, the fish may start to suffocate.

Despite some drawbacks, HEPPs are still viable. Their development in the world continues. The International Energy Agency predicts global HEPP cumulative operating power will increase over the next 5 years by 103 GW – from 1,324 (2020) to 1,427 GW. HEPP produced a large share of the world's electricity in 2020 (16.85%) [1], and the shortcomings of these plants are not so significant, so it is likely that existing power plants will be in operation for a long time and even new ones will be built. The problem of valuable fish spawning grounds can be addressed in the following ways: countries where HEPPs are being built, could leave half of their land for accesses of valuable fish to their spawning grounds, and on the other half of their territory, they should be allowed to operate existing hydropower plants and even build new ones. Tackling the climate crisis is a much more pressing issue than creating good spawning conditions for the valuable species of fishes in the whole territory of the country.

Environment Polluting Power Plants

The most polluting power plants (PPP) currently operating in the global power production sector are based on fossil fuels. Their shares in the total global power production per the year 2020 were the following [1]:

- Coal power plants (CPP) – 33.79%;
- Natural gas power plants (NGPP) – 22.79%;
- Nuclear power plants (NPP) – 10.12%;
- Oil power plants (OPP) – 4.37%.

OPPs are lagging behind in the global power production area. Besides, they are big polluters of the environment (840 g of CO2 eq. per kWh, [13]). Amounts of power production by the OPPs on a global scale are declining as well as power production by the CPPs. However, CPPs currently produce the largest share of electricity worldwide among all technologies of power production. As a result, they fall into the group of the six most important comparable electricity generation technologies as an important player.

Coal Power Plants

For many years, coal was the most important primary source of electricity. At the same time, coal was the most polluting source. The efficiency of this technology is low (33-38%). This means that only about a third of the thermal energy from coal combustion is used to generate electricity, and most of it accounts for the chemical and thermal pollution of environment. The high carbon intensity (1,001-1,240 g of CO_2 eq. per kWh) [13] and high installation costs ($ 3,672-5,861 per kW) [14] of these power plants make them unattractive, so they are decommissioned one after another. The higher carbon intensity coal is the lower quality called lignite or brown coal. The modern power plants, which have a higher efficiency of operation, have higher installation costs.

The installed capacity and electricity production of CPPs operating worldwide is declining every year. Coal mines are also being closed around the world, and solar power plants are often installed in the devastated areas.

CPPs and coal mines are responsible for the reduced life expectancy of millions of people, especially in those parts of the world where there are many powerful power plants of this kind. To reduce the harmful effects of smog on health, people there are advised to wear protective masks on their faces. Of course, various types of vehicles running on fossil fuels also contribute a lot to air pollution in cities and around the main highways. According to the new data of World Health Organisation (WHO), 90% of the world's population are breathing highly polluted air and about 7 million people die every year due to this reason including 4.2 million premature deaths as a result of outdoor air pollution's negative impact [22, 23]. 95% of the global population are breathing air contaminated by a particular matter whose concentration exceeds the limit recommended by the WHO (10 $\mu g/m^3$).

Natural Gas Power Plants

NGPPs are currently the best among all PPP, in terms of their efficiency and lowest environmental pollution. They account for the second-best share of global power production in 2020 (22.79%) [1], have the lowest installation cost ($2,471 per kW) [14], and the lowest carbon intensity (469-645 g of CO2 eq/kWh) [13] among the same group of power plants and the third-best installed power capacity for the year 2020 (40 GW) among power plants of all types. The most important factors that will determine the winners in the power generation technology competition will be economic indicators such as LCOE and investment costs. These indicators show that NGPP has the greatest potential to survive in the competitive battle among all power generation technologies for the longest time. However, electricity generation data shows that global electricity generation in this type of power plant has been steadily declining over the last few years, while electricity production at renewable power plants is increasing at a pace never seen before.

It is also necessary to take into account the fact that the environment is polluted not only by burning fossil fuels in power plants (coal, natural gas, oil products), but also by various means of transport (oil products, natural gas). Large-scale environmental pollution occurs at all stages of fossil fuel preparation and use. Land, water, and air are being polluted. Significant damage is being done to the whole biosphere, ecosystem, climate, infrastructure, and all types of life in contaminated areas – flora, fauna, and humans. This is evidenced by information about the scale of pipeline incidents in the USA from 1986 to 2012, published on May 30, 2013 (data source – US Department of Transportation Pipeline and Hazardous Materials Safety Administration):

- Number of incidents – 7,978;
- Fatalities – 512;
- Injuries – 2,360;
- Property damage – $6,838,019,931.

Unfortunately, environmental pollution is not just about power plants running on fossil fuels. All other energy producers that use fossil fuels in all branches of the economy are also responsible for the ecocide.

There are more signs that the era of NGPP, like all other fossil fuel power plants, is coming to an end. Some banks are no longer willing to finance fossil fuel-based projects. Coal mining and construction of new coal-fired power plants were the first to face this problem. Recently, the president of the European Investment Bank, Dr. Werner Hoyer declared that "gas is over" [24].

People's attitudes towards the decommissioning of fossil fuel power plants are also conducive to major changes and their views are also very important. Most of them support the green course in the energy, transport, and other sectors of the economy, as it brings many new jobs, strengthens hopes that climate change will be tackled, and that it will eventually be possible to breathe cleaner air.

Nuclear Power Plants

A big future for the NPPs was predicted at the beginning of the nuclear energy era. According to a forecast published in 1974, the total power of NPPs was to increase up to 5.3 TW by 2000, but in reality, substantially fewer power plants were operating in the world last year - their total global power was only 0.370 TW [25]. It is a special type of power plant. NPP installation costs ($6,183 per kW) [14] and LCOE ($163 per MWh) [3] currently are the highest among all types of power plants. Besides, the price of decommissioning of the NPP, which can last for a very long time (about 20 years), is too high and not so much different from the installation costs of the NPP. This could still be acceptable if devastating accidents at this type of power plant were not possible. Unfortunately, no one can give a 100% guarantee that accidents are impossible. In addition, today's world has a number of nuclear-armed militant states and many terrorist organizations can choose to target NPP at any time. Therefore, NPPs are presently not attractive worldwide despite their being part of the low-carbon power plants group.

If the era of coal-fired power plants is coming to an end, the same cannot be said of the NPP. Some countries around the world are developing a new type of small modular nuclear reactor for more flexible power generation [26]. The capacity of one small modular reactor (SMR) makes up to 300 MWe. Currently, about 50 SMRs are under design and construction in many countries. The company, NuScale Power based in Portland, Oregon developed SMR whose capacity is 50 MWe [27]. 12 units of SMR can be connected in parallel to obtain a total capacity of 600 MWe. This company is also developing smaller SMRs. Their micropower modules will have capacities from 10 to 50 MWe and heat pipe reactors (for heat production) will have an even smaller capacity – from 1 to 10 MW [28].

However, according to Scientific American, concerns about the high price and safety issues of SMR remain [27]. NuScale's first project of 12SMR x 60 MW (720 MW) was at first valued at about $3 billion in 2017, and now its cost has risen to over $6 billion [27]. Installation costs for this case can be easily calculated – it makes up $8,333 per kW. Meanwhile, the installation cost (or Capex) of PVPP for the year 2020 was $1,248 per kW (Table 1) and a very significant further decline in this index is projected until 2050.

Comparison of the RES-Based and Polluting Power Plants

The comparison of RESPP and polluting power plants was performed by presenting their most important parameters in two tables. The key parameters of the RESPP are submitted in Table 1 and the corresponding parameters of the environment polluting power plants – on the adjacent page in Table 2.

One of the most important indicators for evaluating power generation technologies is LCOE. As the data in Table 1 and Table 2 show, the LCOE for all RESPPs are much better than the same indicators for the PPP.

The indicator of total global operating capacity (TGC) of power plants as well as the global share (%) of power production per 2020 reflects more the results of the development of various types of power plants in the past, rather than the present achievements. The more intensive development of PVPPs and WPPs started only 10-20 years ago, when the development of PPPs listed in Table 2 had been going on for the last 60-100 years. Therefore, PVPPs and WPPs are still lagging behind according to these indicators. But their real development is just beginning, so rapid changes according to these indicators can be expected very soon.

The maximum capacity of one power plant (or farm) is not a very important indicator. However, in this "competition," the RESPPs are presently in the front positions as well.

Indicators of global installed capacity and globally added power for the year 2020 reflect the current dynamic properties of the development of the studied power generation technologies. Table 1 shows an evident superiority of the RESPPs against the PPPs (Table 2). The technical annual potential of RESPPs for power production is compared with the global electric power demand for the year 2018.

Table 1. Key indicators of the RES-based power plants

Parameters	Units	Types of power plants		
		PVPP	WPP	HEPP
LCOE [3]	\$/MWh	37	40	50 [17]
TGC 2020 [18]	TW	0.728	0.734	1.324
TGC 2025 [18]	TW	(PVPP+WPP) 2.349		1.427
Maximum capacity of one power plant/farm	GW	10 [4] N.Territory, Australia	20 [29] Gansu WF, China	22.5 [20] 3 Gorges China
Global installed capacity 2020 [5]	GW	127	111	14.5
Globally added power for 2020	TWh	+84 [19]	+158 [19]	+178 [19]
Gl. share of power production for 2020	%	3.27 [1]	6.15 [1]	16.85 [1]
Installation cost	\$/kW	1,248 [14]	1,846-4,362	2,769 [14]
Carbon intensity	g CO_2 eq/kWh	48 [13]	12 [13]	4 [13]
Water demand [21]	m^3/MWh	0.020-0.098	0	17 [21]
Fuel demand	–	No	No	No
Technical potential	TWh/a	5,000,000 [2]	160,000	15,000
Global demand	TWh/a	22,315 [30] – 23,398 [31] (2018)		
Premature deaths	per TWh	0.02 [32]	0.04 [32]	0.02 [32]

RESPPs also have excellent indicators of the installation cost (capital expenditures or Capex), carbon intensity, and premature deaths rates from power production per TWh [32]. The installation cost of PVPPs and WPPs still have a trend to decrease. PVPPs have a particularly high untapped potential to increase the efficiency of solar irradiance conversion into electricity. According to the reference [33], the installation cost of residential PVPPs is predicted at €826 per kW in 2030, €650 in 2040, and €537 in 2050. The same indicator for commercial PVPPs should decrease to €623 per kW in 2030, €484 in 2040, and €397 in 2050. After that, solar power plants will no longer have strong competitors in the electricity generation sector.

In terms of carbon intensity, PPPs have serious problems in this area. With the exception of NPPs, PPPs can't achieve such carbon intensity values, and their most impressive drawbacks in this area is high environmental pollution and high premature deaths. Mass pollution and huge losses are associated with fossil fuel extraction, transportation, fuel production, distribution, and combustion in power plants, boilers, vehicles etc. Only NPPs have good carbon intensity and low premature death indicators. Therefore they are classified as low carbon power plants and this is their biggest advantage over other PPPs.

Table 2. Key indicators of the polluting power plants

Parameters	Units	Types of power plants		
		NGPP	CPP	NPP
LCOE [3]	$/MWh	59	112	163
TGC 2020 [18]	TW	1.822	2.097	0.370 [25]
TGC 2025 [18]	TW	1.999	2.079	No data
Maximum capacity	GW	5.6 [34] Surgut-2 Russia	6.72 [35] Tuoketuo China	8.212 [36] Kashiwazaki-Kariwa, Japan
Global installed capacity for 2020	GW	40 [19]	0	7 [37]
Globally added power for 2020	TWh	–117 [19]	–519 [19]	–109 [19]
Gl. share of power production for 2020	%	22.79 [1]	33.79 [1]	10.12 [1]
Installation cost	$/kW	2,471 [14]	3,672-5,861	6,183 [14]
Carbon intensity	g CO_2 eq/kWh	469-645 [13]	1,001-1,240	16 [13]
Water demand [21]	m^3/MWh	0.70-4.10	2.32-6.90	1.02-3.20
Fuel demand	–	Yes	Yes	Yes
Technical potential	TWh/a	Limited resources		
Premature deaths	per TWh	2.82 [32]	32.72-24.62	0.07 [32]

All three discussed RESPPs do not require any fuel and this is a very important advantage against all PPPs based on fossil fuels. RESPPs have huge technically available potentials for power production from their primary renewable energy sources. Meanwhile, technically available potentials for power production in PPPs are restricted by the limited resources of the fossil fuels.

As shown in Figure 1, drawn from the data presented in the reference [38], annual global additions of newly installed capacity by the polluting power plants and by the RES-based power plants during the years 2016-2020 again shows the evident and increasing superiority of the RESPPs. According to many forecasts, this trend will continue and will strengthen in favour of the RESPPs.

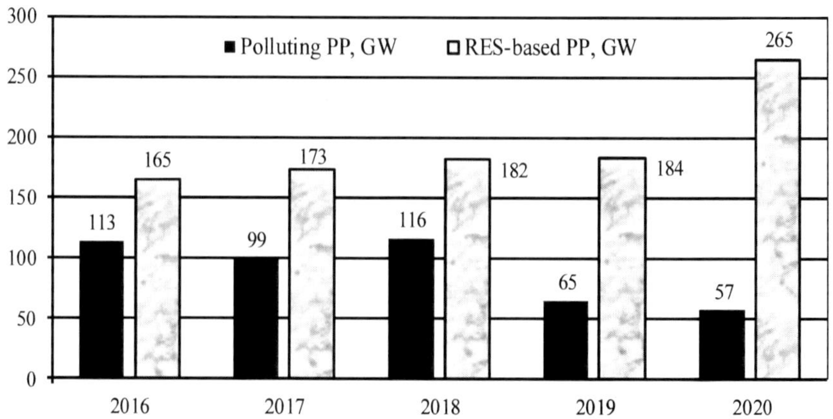

Figure 1. Annual global additions of the polluting power plants and the RES-based power plants capacities during the last five years.

The bar chart (Figure 2) drawn on basis of the Table 1 and Table 2 data shows that total global power production by all operating RESPPs also surged very significantly in the year 2020 (+ 420 TWh). Meanwhile, power production by the globally operating PPPs during the same period experienced a substantial decline during the same year (– 745 TWh).

Figure 3 shows a comparison of the percentage of global energy production by the selected groups of power plants over the last 11 years. The annual global energy production of the NPPs (the red curve) is compared with the energy produced by RESPPs (the green curve). Similarly, the same picture compares global electricity generation in the CPPs (the black curve) and in the LCPPs (low-carbon power plants, the blue curve). Figure 3 is plotted according to the data provided by the reference [1].

As shown in the curves in Figure 3, the volumes of global power production by CPPs and LCPPs leveled off in 2018 and after this the electricity production curve of CPPs started to fall quite rapidly.

The curves representing RESPPs and NPPs never intersected in the past. The smallest difference between their shares of the produced global electric power (1.61%) was in 2001 when the RESPP generated global share was 18.68% and the NPP share 17.07%. [1].

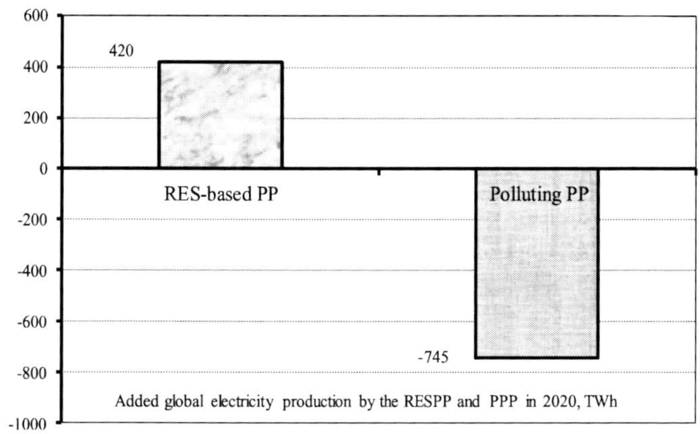

Figure 2. Additions of the globally produced power by the 3 main types of RES-based and the 3 main types of polluting power plants for the year 2020.

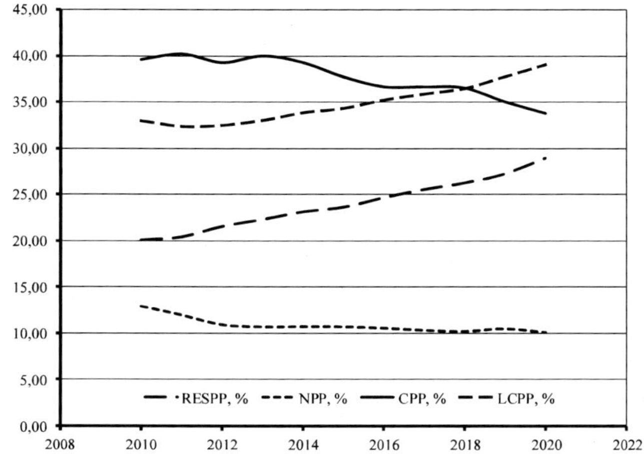

Figure 3. Percentage of global power production by the competing electric power production technologies for years 2010-2020: RESPP, NPP, CPP and LCPP.

A review of a number of sources of information on power generation technologies has led to the finding that the RESPP has the best prospects for successful future development. This conclusion is confirmed by their key power production indicators. The development of solar and wind power plants is currently developing at a pace that few expected at the beginning of this century.

Most interestingly, some researchers and their students working in the field of RESPPs believe that solar energy is currently significantly underestimated. Christian Breyer, a professor of solar economics at the Lappeenranta University of Technology, and his colleagues have corrected their views on the share of solar power plants in the global electricity mix by 2050 [33].

They forecast the following indicators for the future:

- Installation cost (Capex) for utility-scale projects – €246/kW;
- Installation cost (Capex) for residential projects – €537/kW;
- The share of PVPPs in the global electricity mix – from 68% to 76%.

There is a good chance that they may be right, as the technologies of PVPPs are far from exhausting their full potential. Efforts will continue to increase the efficiency of solar cells and modules and reduce the cost of all other equipment.

Other Advantages of the RES-Based Power Plants

The benefits of RESPP are not limited by the good indicators of these power plants. Centralized power generation in power plants of high and very high capacity, transmission, and distribution of the produced power to consumers is still common in many countries around the world. The energy produced by power plants of high capacity only reach consumers after a long way through high-voltage transformer substations, switchyards, power transmission networks, low-voltage transformer substations, and power distribution networks. Electric current, traveling to all corners of the country, creates a sufficiently significant power loss in such an energy system, which can reach up to 30%. This means that about 30% of electricity is generated solely to be wasted on the way from powerful power plants to the consumers.

RES-based power plants, and in particular PVPPs, have superiority in this respect. Using them, it is not difficult to create a distributed electricity generation system, where the prosumer generates electricity itself in its solar power plant and consumes it itself at its production site. In this way, power losses associated with long-distance power supply could be avoided.

Currently, in many countries, buildings are supplied with energy from centralized networks using three inputs: electricity, thermal energy, and natural gas. While a distributed energy generation method is used, buildings can be fully self-sufficient in electricity and heat. In addition, with the rapid decline in battery prices and the popularity of electric vehicles, buildings can also supply power to the electric vehicles of their residents. Microgrids, consisting of PVPPs, WPPs, heat pumps, and batteries, can provide electricity and heat to buildings and electric vehicles of their inhabitants in areas with sufficient wind energy resources without the use of any fuel [39, 40].

The world already has apartment buildings that produce more energy per year than they can consume [41]. Four apartment buildings designed by the solar architect Rolf Disch are being built in Schallstadt, Black Forest, Germany. Electricity supply to the buildings is provided from the local mutual PVPP. The capacity of the plant is 470 kW. The heating system of all buildings is based on the heat pumps "water to water," which uses the heat from the large sewage pipeline. The temperature of this heat source is between 15 and 17°C throughout the year. Therefore, the performance factor of the heat pumps is high (4.5) and electricity consumption by the heat pumps is low. As a result, the surplus of clean electricity produced in the site per year is over 200,000 kWh. This amount of electricity is used for the charging of batteries of local electric vehicles from the apartment buildings.

When the energy system is dominated by very powerful power plants, the supply of energy can be seriously jeopardized in the event of a terrorist attack or war, where enemies want to destroy the country's energy grid and economy for a very long time. A centralized energy system, dominated by only a few high-capacity power plants, makes this harmful task much easier for them. It would be much more difficult for aggressors to destroy a huge number of energy generation systems in a country where most buildings would meet the needs of their energy consumers (including electric vehicles) from their local small-scale RES-based microgrids.

When the principle of distributed electricity generation is applied, a new PVPP can be installed much faster than any other power plant or power supply could be restored in case of failure. There is no need to have highly qualified specialists to install and maintain this power plant. This work can be done by every trained electrician. It would be much more difficult to install and maintain a fossil fuel power plant of any type. There are no moving parts in the solar power plants, no fuel, grease, water, or noise.

Energy raw materials and other minerals often lead to wars, various conflicts, overturns, diversions and provocations. Therefore the whole world is expected to be much more peaceful, the air and the whole environment much cleaner, and the people of planet Earth healthier when the transition to green technologies is implemented in energy, transport, agriculture, and all other sectors of the global economy.

A review of sources of information on the topic of power generation technologies has made it possible to make sure that much of the global energy sector is on the right track. The fastest possible transition to the use of significantly more efficient and cleaner energy production technologies, together with the conservation of the energy produced, would be fully beneficial as it would contribute to solving the most important problems of humanity such as the climate crisis, pollution of the environment, health, and mass extinction of species.

Rapid integration of the new RES-based power plants into the present electric power systems will require a lot of novel scientific knowledge obtained in spheres of renewable power generation, sustainability, artificial intelligence, intelligent computing, and optimal control of electric power flows. This knowledge will be necessary for upgrading the existing power system and making it stable, reliable and cost-effective in the scale of the entire country. The latest scientific knowledge presented at global scientific conferences and published in editable scientific books can be useful for solving the current challenges of modern heat and power engineering [42-44].

CONCLUSION

Typically, creative people specialize in one area to better delve into the issues at hand. However, from time to time it is useful to find out how other professionals solve similar problems and what results they have achieved. Overviews of the key technical, economic, and environmental performance of different technologies used for the same purpose helps to reveal the advantages and disadvantages of those technologies. The results of the reviews can help decision-makers compare the most effective technologies according to the key indicators and to select the best and most promising of them.

This review of the presently used main electric power generation technologies has shown that currently, RES-based power plants are the most efficient in terms of many indicators: LCOE, installation cost (Capex), carbon intensity, development rates, premature mortality of the related persons, and more. Solar power plants have the best forecast for the future. The share of solar power plants in the global electricity mix is forecast to be from 68% to 76% by 2050.

The review showed that currently RES-based electricity technologies have made it possible to address energy generation and supply challenges in a much more technical, environmentally friendly and cost-effective manner (often even several times cheaper) than they were at the beginning of this century. Therefore, it is worth going further along this path.

Acknowledgments

The author is very grateful to Professor Pandian Vasant, Universiti Teknologi Petronas, Malaysia, for the proposal to write a chapter for this book and to all editors for the remarks and useful advice. Your attention and prompt communication have greatly facilitated this work.

References

[1] Ritchie H. Electricity Mix. *Our World in Data*. Online [04.05.2021] Available at: https://ourworldindata.org/electricity-mix.

[2] *World Energy Assessment: Energy and the Challenge of Sustainability*. United Nations, UNDP. 2000, 506 p. Online [04.05.2021]. Available at: https://sustainabledevelopment.un.org/index.php?page=view&type=400&nr=2423&menu=35.

[3] Carroll D. Biden plans to fight climate change in a way no U.S. president has done before. *Energy Matters to Climate Change*. January 20, 2021. Online [23.04.2021]. Available at: https://www.e-mc2.gr/ el/news/biden-plans-fight-climate-change-way-no-us-president-has-done.

[4] Carroll D. Major Milestone for world's biggest solar project. *PV Magazine*. January 29, 2021. Online [25.04.2021]. Available online at: https://www.pv-magazine.com/2021/01/29/major-milestone-for-worlds-biggest-solar-project/.

[5] *World Adds Record New Renewable Energy Capacity in 2020*. IRENA press release. April 05, 2021. Online [25.04.2021]. Available at: https://irena.org/newsroom/pressreleases/2021/Apr/ World-Adds-Record-New-Renewable-Energy-Capacity-in-2020.

[6] Goetzberger A., Zastrow A. *International Journal of Solar Energy.* On the Coexistence of Solar-Energy Conversion and Plant Cultivation. 1982, vol. 1, is. 1, pp. 55-69. Available at: http://DOI:10. 1080/01425918208909875.

[7] Agrophotovoltaics: *High Harvesting Yield in Hot Summer of 2018.* Fraunhofer Institute for Solar Energy Systems. Press Release. April 12, 2019. Online [27.04.2021]. Available at: https://www.ise. fraunhofer.de/en/press-media/press-releases/2019/agrophoto voltaics-hight-harvesting-yield-in-hot-summer-of-2018.html.

[8] Andrew A. C., Higgins Ch. W., Smallman M. A. etc. *Front. Sustain. Food Syst.* Herbage Yield, Lamb Growth and Foraging Behavior in Agrivoltaic Production System. April 29, 2021. Available at: https://doi.org/10.3389/fsufs.2021.659175.

[9] Schindele S., Trommsdorff M., Schlaak A. etc. *Applied Energy.* Implementation of agrophotovoltaics: Techno-economic analysis of the price-performance ratio and its policy implications. 2020, vol. 265, 19 p. Available at: https://doi.org/10.1016/j.apenergy. 2020. 114737.

[10] Famoso F., Lanzafame R., Maenza S., Scandura P. F. *Energy Procedia. Performance Comparison between Micro-inverter and String-inverter Photovoltaic Systems.* 2015, vol. 81, pp. 526-539. Available at: https://www.researchgate.net/publication/290050546 _Performance_Comparison_between_Micro-inverter_and_String-inverter_Photovoltaic_Systems.

[11] Ingram E. Global wind industry to deploy 1 TW of new capacity through 2030. *Renewable Energy World.* April 8, 2021. Online [28.04.2021]. Available at: https://www.renewableenergyworld. com/wind-power/global-wind-industry-to-deploy-1-tw-of-new-capacity-through-2030/#gref.

[12] Hall M. Solar and wind together overhaul global hydro capacity. *PV Magazine – Australia.* April 8, 2021. Online [29.04.2021]. Available at: https://www.pv-magazine-australia.com/2021/04/08/solar-and-wind-together-overhaul-global-hydro-capacity/.

[13] Adomavicius V., Kaminickas M. *Energetika. Analysis of PV Power Plants Future Development Possibilities.* 2014, vol. 60, issue 4, pp. 233-248. https://www.researchgate.net/publication/271836686_Analysis_of_PV_power_future_development_possibilities.

[14] Base overnight costs of new electricity generating stations in the United States in 2020, by major technology. *Company Statista. Energy and Environment*. Online [29.04.2021]. https://www.statista.com/statistics/519118/power-plant-base-overnight-costs-in-the-us-by-technology/.

[15] Sovacool B. K. *Energy Policy*. Contextualizing avian mortality: A preliminary appraisal of bird and bat fatalities from wind, fossil-fuel, and nuclear electricity. 2009, v. 37, pp. 2241–2248. Available at: https://doi:10.1016/j.enpol.2009.02.011.

[16] Wang Sh., Wang S., Smith P. *Renewable and Sustainable Energy Reviews.* Ecological impacts of wind farms on birds: Questions, hypotheses, and research needs. 2015, vol. 44, pp. 599-607. Available at: http://dx.doi.org/10.1016/j.rser.2015.01.031.

[17] Levelized cost of electricity (LCOE) of hydropower worldwide from 2010 to 2019. Company Statista. *Energy and Environment*. Online [03.05.2021]. Available at: https://www.statista.com/statistics/799349/lcoe-of-hydropower-worldwide/.

[18] Analysis and forecast to 2025. *Report Renewables 2020*. Paris, International Energy Agency. November 2020, 172 p. Online [04.05.2021]. Available at: https://www.iea.org/reports/renewables-2020.

[19] *Global overview: Capacity, supply and emissions*. IEA, Paris, 2020. Online [04.05.2021]. Available at: https://www.iea.org/reports/electricity-market-report-december-2020/2020-global-overview-capacity-supply-and-emissions.

[20] World's biggest hydroelectric power plants. *Power Technology*. March 05, 2021. Online [05.05.2021]. Available at: https://www.power-technology.com/features/worlds-biggest-hydroelectric-power-plants/.

[21] McMahon J. E., Price S. K. *Water and Energy Interactions Lawrence Berkeley National Laboratory.* August 2011, 40 p.

[22] *Air pollution.* World Health Organization (WHO). Online [04.05.2021]. Available at: https://www.who.int/health-topics/air-pollution#tab=tab_1.

[23] Ritchie H., Roser M. *Outdoor Air Pollution.* OurWorldInData. Online [07.05.2021]. Available at: https://ourworldindata.org/outdoor-air-pollution.

[24] Queally J. Climate Leaders Celebrate as European Investment Bank Chief Declares 'Gas Is Over.' *EcoWatch.* Online [28.04.2021]. Available at: https://www.ecowatch.com/european-investment-bank-fossil-fuels-2650067861.html?rebelltitem=4#rebelltitem4.

[25] Schneider J., Froggatt A., Hareman J. etc. *The World Nuclear Industry Status Report.* Paris, September 2019, 323 p. Online [10.05.2021]. Available at: https://www.researchgate.net/publication/336870789_The_World_Nuclear_Industry_Status_Report_2019/link/5db7f69f92851c8180121201/download.

[26] *Small modular reactors: flexible and affordable power generation.* International Atomic Energy Agency (IAEA). Online [27.04.2021]. Available at: https://www.iaea.org/topics/small-modular-reactors.

[27] Levitan D. *First U.S. Small Nuclear Reactor Design Is Approved.* Scientific American. September 9, 2020. Online [28.04.2021]. Available at: https://www.scientificamerican.com/article/first-u-s-small-nuclear-reactor-design-is-approved/.

[28] Conca J. *NuScale Small Modular Nuclear Reactor Moves another Step Forward.* Forbes. January 14, 2021. Online [02.05.2021]. Available at: https://www.forbes.com/sites/jamesconca/2021/01/14/nuscale-small-modular-nuclear-reactor-moves-another-step-forward/ ?sh=320c82265264.

[29] Vyas K. The 11+ Biggest Wind Farms and Wind Power Constructions That Reduce Carbon Footprint. *Science.* February 15, 2018. Online [08.05.2021]. Available at: https://interesting engineering.com/the-11-biggest-wind-farms-and-wind-power-constructions-that-reduce-carbon-footprint.

[30] *Electricity Information: Overview*. International Energy Agency. Report. July 2020. Online [09.05.2021]. Available at: https://www.iea.org/reports/electricity-information-overview.

[31] *Net consumption of electricity worldwide in select years from 1980 to 2018*. Company Statista. Online [10.05.2021]. Available at: https://www.statista.com/statistics/280704/world-power-consumption/.

[32] Ritchie H. Nuclear energy and renewables are far, far safer than fossil fuels. *Death Rates from Energy Production per TWh*. Online [10.05.2021] Available at: https://ourworldindata.org/nuclear-energy.

[33] Bellini E. Solar may cover 75% of global electricity demand by 2050. *PV Magazine*. April 13, 2021. Online [10.05.2021]. Available at: https://www.pv-magazine.com/2021/04/13/solar-may-cover-75-of-global-electricity-demand-by-2050/.

[34] Profiling five of the world's biggest gas-fired power stations. Surgut-2, Russian Federation. *NS Energy*. November 05, 2019. Online [10.05.2021]. Available at: https://www.nsenergybusiness.com/features/worlds-biggest-gas-fired-power-plants/#.

[35] *Datang Tuoketuo Coal Power Station*. Global Energy Observatory. Online [10.05.2021]. Available at: http://globalenergyobservatory.org/geoid/41851.

[36] *Kashiwazaki-Kariwa Nuclear Power Station*, Japan. Online [04.05.2021]. Available at: https://www.power-technology.com/projects/kashiwazaki/.

[37] *Global nuclear capacity by scenario, 2000-2040*. International Energy Agency. Online [11.05.2021]. Available at: https://www.iea.org/reports/nuclear-power.

[38] Masterson V. 5 Milestones in Green Energy. *World Economic Forum*. April 14, 2021. Online [10.05.2021]. Available at: https://www.weforum.org/agenda/2021/04/renewables-record-capacity-solar-wind-nuclear/.

[39] Adomavičius V., Valickas J., Petrauskas G. ect. *Proceedings of 6th International Conference Trends in Agriculture Engineering.*

Potential of village house for sustainable energy production. 2016, Prague, pp. 17-25. Available at: https://www.researchgate.net/publication/308305426_POTENTIAL_OF_VILLAGE_HOUSE_FOR_SUSTAINABLE_ENERGY_PRODUCTION.

[40] Adomavicius V., Valickas J., Petrauskas G. etc. *Proceedings of 11th International Conference Intelligent Technologies in Logistics and Mechatronics Systems.* Concept of RES-based microgrid for apartment building. April 28-29, 2016, Panevezys faculty at KTU, Lithuania, pp. 1-6. Available at: https://www.researchgate.net/publication/309430727_Concept_of_RES-based_microgrid_for_apartment_building.

[41] *Climate Houses in the Black Forest Show Architecture of the Future.* PV Europe. March 30, 2021. Online [12.05.2021]. Available at: https://www.pveurope.eu/installation/energy-efficiency-climate-houses-black-forest-show-architecture-future.

[42] *Proceedings of the 1st International Conference on Intelligent Computing and Optimization (ICO'2018).* Editors: Vasant P., Zelinka I., Weber G. W. (Eds.) Available at: https://www.springer.com/ gp/book/9783030009786.

[43] *Proceedings of the 2nd International Conference on Intelligent Computing and Optimization 2019 (ICO 2019).* Editors: Vasant P., Zelinka I., Weber G. W. (Eds.) Available at: https://www.springer.com/gp/book/9783030335847.

[44] *Proceedings of the 3rd International Conference on Intelligent Computing and Optimization* 2020 (ICO 2020). Editors: Vasant P., Zelinka I., Weber G. W. (Eds.) Available at: https://link.springer.com /book/10.1007/978-3-030-68154-8.

In: Advances of Machine Learning … ISBN: 978-1-68507-211-7
Editors: Pandian Vasant et al.

Chapter 7

A COMPREHENSIVE STUDY OF SYSTEM BUILDING BLOCKS FOR RADIO FREQUENCY ENERGY HARVESTING

***Bhuvnesh Khantwal**, *Reeta Verma*† and *Paras*‡**
Department of E & CE College of Technology,
G. B. Pant University of Agriculture and Technology,
Pantnagar, Uttarakhand, India

ABSTRACT

Our environment has been condensing by various radio frequencies (RF) waves because of a rapid increase in the use of wireless communication technologies over the last few decades. In fact, we are living in a sea of RF radiated energy all around us. This factor accompanied by the advancing electronics has motivated researchers to extend the idea of energy harvesting to include RF harvesting. This technology is one of the major keys to replace conventional battery sources present in low-powered devices, helping reduce battery replacement costs and contributing to essentially a greener production.

* Corresponding Author's Email: bhuvkwal@gmail.com.
† Corresponding Author's Email: reetaverma82@gmail.com.
‡ Corresponding Author's Email: paras.ec.pant@gmail.com.

This paper presents a comprehensive study of the RF energy harvesting (RFEH) system. The basic building blocks of a typical RF energy harvester are antenna, matching circuit, rectifier, and power management unit. Each of these blocks along with various circuit topologies involved is discussed in an orderly manner. This study gives a broad view of the current trends and recent developments in this domain. Critical design considerations have been pointed out in all the system blocks for efficient power conversion. Furthermore, the major challenges that need to be addressed in future researches are also highlighted.

Keywords: RFEH, antenna, matching, rectifier, power management, efficiency

1. Introduction

The concept of wireless power transfer is not new. The first attempt in this series can be dated back to the late 1800s. The great scientist Nicola Tesla was the first one who demonstrated wireless power transmission using the famous Tesla Coil. In the early 1900s, he made a giant Tesla coil for mass transfer of wireless power in New York. Though it could not be further expanded to a large scale, it sparked the possibilities regarding wireless power transmission. Further experiments have been conducted since then to demonstrate various practical possibilities of wireless power transfer. In 1960 authors in [1] designed a small helicopter that was powered wirelessly by a 270W, 2.45GHz source at the ground and was able to fly up to 50 feet high. A high power space-to-earth power transfer system was proposed in [2]. It used enormous transmitting antennas at the satellite station to transmit 2.7 GW of power at a distance of 36000km with 45 percent conversion efficiency.

Initially, wireless power and information transfer were seen as separate entities. Wireless Information Transfer (WIT) systems were optimized with a view of maximizing the information content whereas Wireless Power Transfer (WPT) systems were designed for maximizing power transferability. But over recent years, design objectives of WIT and WPT systems are being aligned. This has raised the hopes of using a single wireless transfer for both information and energy harvesting. Simultaneous wireless information and power transfer (SWIPT) [3, 23] has been proposed in this direction for providing power especially to low power areas such as remotely deployed wireless sensor networks (WSNs). However, both the information and power in transmitting waves cannot be increased at the same time; there is a trade-off between these two. This calls for a re-designing of the existing RF transfer systems for achieving the best of both worlds [4]. Currently, the co-design of WIT and WPT systems is in its early stages. This field has a lot of hurdles to attain a state of maturity and will take some time before it can grow to the point of practical utilization. However, harvesting energy from the already available RF WIT systems has drawn much attention in the last two decades.

Most of the wireless communication technologies work in the radio frequency band. This band falls under the category of non-ionizing radiations. RF waves have frequencies ranging from around 100 kHz to 300GHz. Though not as hazardous as ionizing radiations, non-Ionizing radiations too have their harmful impacts on health if used for a prolonged time [5]. Apart from having a heating effect on eyes and other body tissues, they are a potential carcinogenic source. Hence, there are several standards for the location and maximum power transmission of RF towers for the safeguard of public health. Federal Communications Commission (FCC) is responsible for regulating various RF communications in the USA. Guidelines for the maximum exposure limits to RF radiations have been set by the international commission on non-ionizing radiation protection (ICNIRP) under the world health organization (WHO). Legal values vary from country to country. For example, India has at least ten times more stringent standards than countries like the USA, Canada, and Japan despite all of them adopting ICNIRP guidelines. At 1800 MHz, India has an exposure limit of $0.9 W/m^2$ as compared to $12\ W/m^2$ for the other three countries [112].

RF sources can be categorized as dedicated and ambient sources [4]. Over the past years, Dedicated RF sources have been commercialized. These sources are dedicated to providing reliable wireless power to low-energy systems. These sources are not 'free' and have power constraint norms. Due to this, they have a limited power range. Various such dedicated sources may be needed for powering a wireless network. For these reasons, these systems have not yet received widespread acceptance. Radio frequencies (RF), on the other hand, are present more or less everywhere and are completely free to be converted into usable power. Hence this widespread availability of ambient RFs has provided an impetus for research in ambient RF harvesting.

Several practical applications have been contributed intensively towards the motivation for research on (ambient) RF harvesting. WSN has become one of the most important systems exploiting this energy. A WSN consists of several sensor nodes which can communicate with others wirelessly. Each sensor node requires a separate power source for its operation. These power sources need to be recharged at regular intervals. This may be troublesome and costly, especially in the situation when the sensor nodes are not easily accessible. Considering the aforementioned facts and low power requirements of WSNs, RF waves can be a good alternative to the conventional power sources used in sensor nodes. For instance, authors in [6, 40] designed an RF-powered node capable of transmitting the signal at an average of 5kb/sec with an uplink/downlink frequency of 2.45GHz/915MHz. It could work with an input threshold power level of -17.1dBm and transmit power of -12.5dBm.

Radio-frequency identification (RFID) is another major field of application of RF energy harvesting. These systems are used for tracking and identification purposes. RFID tags are used at several places such as logistics and supply chain management, accurate race timings, book identification in libraries, etc. Earlier the RFID devices were mostly passive devices that worked by reflecting a part of the received power back to the reader. With the advent of RF harvesting in RFID, these devices can perform several other tasks rather than just being a simple RF reflector. This has led to the development of smart tags which have intelligent power management, on-chip processing of data, and can actively communicate with the reader [4]. In addition to that, several portable and low-power devices like headsets and digital watches can also be powered using a well-designed RFEH system.

Apart from ambient RF radiations, solar, thermal, and vibrations are other alternate low energy sources. Each of them has its advantages and disadvantages based on their availability, energy densities, efficiency, and cost as shown in table [51, 108, 109]. Solar energy is the most mature, studied [146], and commercialized technology among all of these. Thermal energy harvesting is based on the Seebeck effect which converts thermal energy to electrical energy by utilizing temperature gradient. A thermoelectric device doing this is called a thermoelectric generator. The reverse process, i.e., electrical to thermal energy conversion, is also possible by using another thermoelectric device deploying the Peltier effect. Thermoelectric generators can be deployed either to harvest energy from freely available environmental temperatures or heat produced from machines. Some commercialization of vibrational energy has also been done but it is mostly limited to high-value applications because of its high cost and low efficiency. Vibrational energy can be harvested by three techniques; these are piezoelectric, electrostatic, and electromagnetic. Piezoelectric materials can convert applied mechanical strain into electrical energy. Harvesting using the piezoelectric effect provides the advantage of miniaturization and simple structure. The electrostatic generation method is based on capacitance variation due to vibrations. This generation method is compatible with Micro-Electro-Mechanical Systems (MEMS) but has the disadvantage of lower energy density. The electromagnetic generation method, deploying coil and magnet arrangement, has high energy density but this method may not be viable for small spaces. Vibrational energy can be harvested from human body motions, train track vibrations, sea waves, etc. The major challenge in vibrational energy is the variation of source vibration frequency which may deviate largely from the resonant frequency of the generator.

Though RF radiations lie in the lower strata when it comes to power density, it becomes a potential candidate for many applications because of its boundless availability, whether indoors or outdoors and because of its perpetual nature, be it day or night. However, apart from low power, the unreliable power level is another major setback of RFEH systems. Also, the harvesting systems have a problem of low power conversion efficiency. Intensive researches have been done in the past 2 decades for improving upon these challenges. The researchers try to design systems that can absorb more and more energy from the environment as well as dissipate lesser and lesser energy in the harvesting system.

Table 1. Comparison of some ambient power sources

Source	Harvesting Method	Power Destiny (Typical values)	Advantages	Disadvantages
Solar	Solar panels/cells	Indoors 10 $\mu W/cm^2$ Outdoors: 10 mW/cm^2	High power density, Mature	Not always available. Not implantable Expensive
Vibration	Electrostatic, piezoelectric or electromagnetic	Human:4$\mu W/cm^2$ Industrial:1-10 mW/cm^2	Implantable, High efficiency	Not always available Physical limitation of material
Thermal	Thermoelectric (Seebeck effect)	Human: 30 $\mu W/cm^2$ Industrial:1- 10 mW/cm^2	High power destiny, Implantable	Not always available Excess waste heat
RF	Antenna	0.1 μW– 1 mW/cm^2	Always available, Implantable	Low Density, Low efficiency of harvesting
Airflow	Generator	200-500 $\mu W/cm^2$	Not suitable for indoors	High power density

Because of the popularization of this exciting energy source, there have been many pieces of research focusing on its different building blocks. Many reviews on the subject have been presented in the past. Table 2 shows a list of reference works for an RF harvesting system. There seems a need to present the complete RF harvesting system with all components and aspects of the subject threaded together in a single paper, putting all the basic and necessary details intact in a single place. Keeping that in mind, this work presents a complete and thorough overview of the RF harvesting system starting from the RF radiations in the environment to the final load device. The paper has been written with a motive for researchers to get a big picture of this exciting technology, examine the system as a whole, and know about what different design techniques are already available in particular sections of the system. Many design aspects of a typical RF harvesting system have been touched on in this study. A good theoretical background associated with each of the system blocks is also presented. Furthermore, design strategies are presented in every section for getting a wide view of the design considerations for better reception and conversion efficiency. The system designs and the corresponding analyses presented show the essence of compatibility between different system blocks for achieving higher efficiencies. The rest of this paper is organized as follows: Section 2 gives an overview of the complete system along with an analysis of RF radiations in the environment. The receiving antenna of an RF harvester, its characteristics, and various design improvements are discussed in section 3. Section 4 addresses the problem of matching the antenna to the rest of the circuit with help of a matching circuit. Rectification of the AC signal, received from an antenna, into DC signal is discussed under section 5. Section 6 provides various ways to feed the load and store the energy provided by the rectifier stage. Section 7 discusses the essence of Power management in an RF energy harvester. Major challenges in RF harvesting and some important design considerations are outlined in section 8 and section 9 gives the conclusion of the paper.

Table 2. Reference works on RF harvesting technology

Reference	Aspects Discussed
[4]	Architecture
[125, 122]	Antenna Characteristics and Design
[125, 131 107, 132]	Antenna and Matching
[125, 126, 130, 133]	Rectifier Design
[127, 128, 129]	Power Management
[12]	Efficiency

2. Overview of an RFEH System

A typical RFEH system is shown in Figure 1. The radio signal power is transmitted from the source antenna at the transmitter side. The transmitted signal undergoes losses during its propagation through transmission media. At the receiver end, the receiving antenna receives the RF waves corresponding to some specifically tuned frequencies. The RF power is converted into electrical power by the receiving antenna. This electrical power then goes to an impedance matching circuit which prevents reflection due to impedance mismatch between antenna and rectifier stages. The rectifier circuit converts the received alternating RF signal to a DC voltage level. This voltage level produced by the rectifier may not be enough to be used for practical application. A voltage multiplier may be used to step up the voltage. The output power from the rectifier/voltage multiplier is then used to charge a storage bank through a power management unit (PMU).

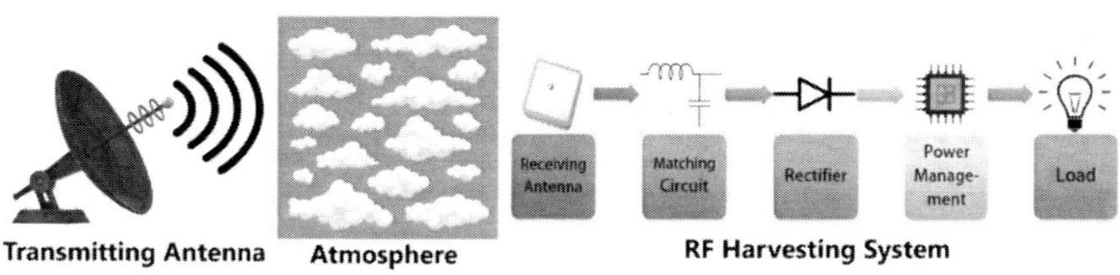

Figure 1. A typical RFEH system block diagram.

Some basic parameters of an RFEH system are as follows.

2.1. Power Conversion Efficiency

The overall power conversion efficiency (PCE) of the system depends on the power conversion efficiency of each system block. These blocks having a major contribution to PCE are antenna, rectifier, matching circuit, and load or storage device. The overall efficiency of the system $PCE_{overall}$ can be defined as given in equation (1):

$$PCE_{overall} = \frac{OutputDCpower}{InputRFpower} = PCE_{Antenna} . PCE_{Matching} . PCE_{Rectifier} . PCE_{Load\ or\ Storage} \quad (1)$$

Each of the blocks must be highly efficient for an acceptable power output level. The major cause of power loss in RFEH systems is parasitic effects, impedance mismatch, threshold voltage, and leakage currents in semiconductor devices. In the text that follows we discuss more on these losses.

2.2. Sensitivity

Sensitivity is an important parameter of RFEH systems. It characterizes the minimum power level of the RF signal for which the system can work desirably. It may be defined in the convenient units of dBm as shown in equation (2) [51]:

$$Sensitivity\ (dBm) = 10\ log_{10}\left(\frac{P_{min}}{1\ mW}\right) \quad (2)$$

Lower is the sensitivity; better is the performance in low power environments. This lets the system harvest energy in the farthest of locations where the power levels are too feeble. RF harvesting systems with sensitivity as low as -30dBm have been designed; for example-25 dBm of sensitivity achieved in [7].

2.3. Frequency Bands and Availability

An RF energy harvesting system should be designed by keeping in mind the frequency spectrum that is available in the environment. One technology may have more power density than others at a particular place. The system has to be optimized and tuned so that the maximal amount of power can be retrieved. The magnitude of this spectrum is dependent on the distance from the source and losses present in the environment [51]. Hence a sound knowledge of these factors becomes important before the design of the energy harvester.

Generally, a large amount of the transmitted RF power lies in VHF (30-300MHz) and UHF (300MHz-3GHz) frequency bands which makes them suitable for energy harvesting applications. Below the VHF band, the frequencies used are for ground wave propagation. They require the antennas to be large, linearly polarized and positioned close to the earth. Hence they are not much suitable for RFEH systems. Digital mobile communications such as Bluetooth, Wi-Fi, 3G, and GSM lie in the UHF band. The power contained in these frequencies is user-dependent. More number of users increases the power intensity of these bands. Figure 2 show some applications of RF harvesting along with the corresponding typical power levels required for RF harvesting for them.

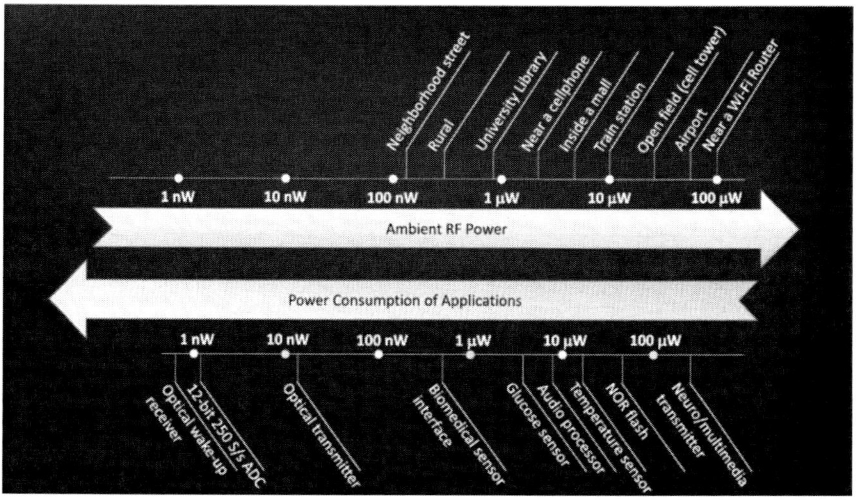

Figure 2. RF Power level requirement of some applications and Typical RF power level at various locations [115, 116, 117, 118].

The TV and FM transmission has a constant power level, independent of the number of users. TV transmission is one of the richest sources of RF power at most locations. Different countries may have different frequency bands for TV and FM broadcasting. Table 3 shows some communication methods and their respective frequency ranges. The values are based on The National Frequency Allocation Plan (NFAP) data [99].

Table 3. RF technologies and their frequency bands

S.No.	Technology	Frequency range
1	TV	585-698 MHz
2	FM	87.5 - 108 MHz
3	GSM900	806-960 MHz
4	GSM1800	1710- 1930 MHz
5	Bluetooth, Wi-Fi	2.4 MHz ISM band

Many Broadcastings and cell phone services are provided using space waves. These waves travel in the troposphere. There is a maximum distance between harvesting and transmitting antenna up to which both can see each other, depicted in Figure 3. This distance is called line of sight (LOS) distance and can be approximately given by equation (3) [100]:

$$d_{los}(metres) = \sqrt{2R}\left(\sqrt{h_t} + \sqrt{h_r}\right) \quad (3)$$

where R is the radius of the earth, h_t *and* h_r are heights of transmitting and receiving antennas respectively (all in meters). This distance is a rough measure of distances up to which a system can harvest the slightest energy from a particular station.

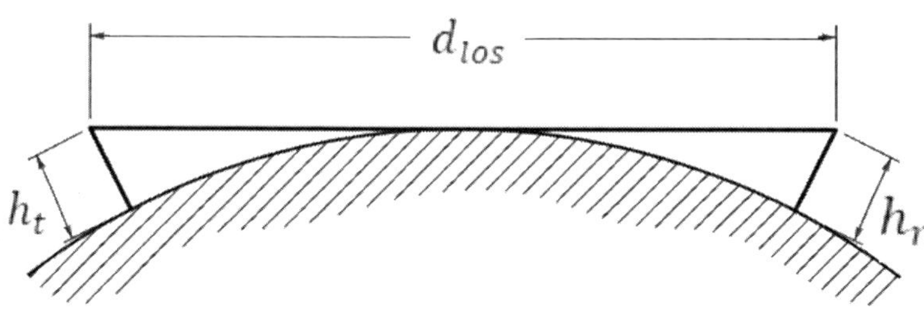

Figure 3. Line of sight distance.

2.4. Near and Far-Field Radiations

Three types of fields are generated by a transmitting antenna. These are known as the radiation field, Induction field, and electrostatic field. All three of these fields die differently with distance. The Induction and electrostatic fields have r^{-2} and r^{-3} variation respectively with distance hence they die rapidly. These two fields are mostly limited to thc so-called near field of an antenna. The near field is the spherical region around the antenna with a radius r [12] as shown in Figure 4.

$$r = \frac{2L^2}{\lambda} \tag{4}$$

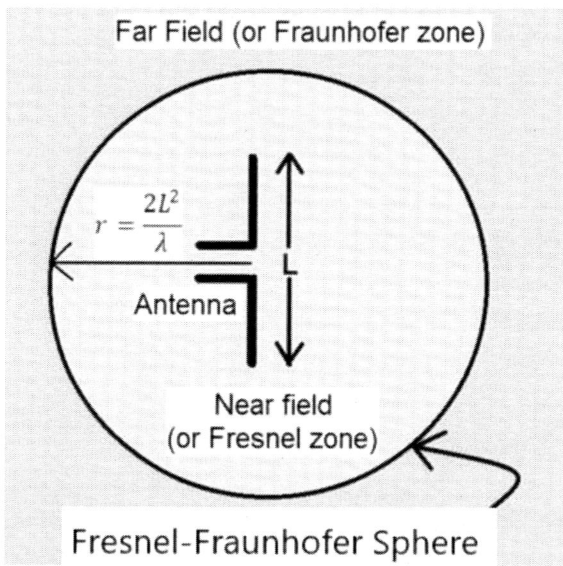

Figure 4. Near and far fields.

where L is the largest dimension of the antenna. As we move away from the antenna in the far-field, the only component remaining is the radiation field which essentially has the characteristics of an EM wave. For an EM wave, the power is proportional to the square of the field amplitude. Therefore in the far-field, the RF waves die inversely as the square of the distance. For a transmitted power P_t, the power received by a receiving antenna in far-field is given by Friis transmission equation as [101]:

$$P_r = \frac{P_t A_t A_r}{\lambda^2 d^2} \tag{5}$$

where d is the distance between transmitting and receiving antenna, $\lambda$is the wavelength of RF waves, A_t and A_r are the effective apertures of transmitting and receiving antenna respectively.

In the near field, energy harvesting is based on inductive coupling [8] and magnetic resonance coupling [9]. These methods require good calibration and positioning of the receiver. Additionally, electric and magnetic fields in this region have a very complex behavior as compared to far-field EM waves. Hence estimation of power is very difficult in this region. Passive RFID and cell phone charging are examples of near-field energy harvesting systems. Because the input power and efficiency of power conversion are more in the near field of the antenna [10-12] and the losses are more in far-field radiations, we see most of the commercial WPT systems employing near field harvesting. For example, a wireless mobile phone charger. However, in practical scenarios of our interest, wireless harvesting systems are located far away from the radiating source. far-field RF waves have typical power densities ranging from 0.2nW/cm^2 to 1uW/cm^2 [12]. Hence it becomes important to customize the RF harvester to perform well in these low-power far-field regions. All blocks of the RFEH system must be designed and optimized for maximally RF energy harvesting. This requires the knowledge of the distance from the source, wave polarization, direction of maximum received power, bands of frequency in the environment, and fluctuations in these parameters. Each block of the RFEH system is discussed in detail in the following sections keeping in mind the aforementioned parameters.

3. Antenna

3.1. Parameters of Interest

An antenna is a fundamental part of the RFEH system. It is a metallic structure that converts RF radiations into electrical energy. It can be of various shapes and sizes depending on the application requirement. Apart from physical parameters of shape and size, an antenna has several other important characteristics. These characteristics are broadly divided into input characteristics and radiation characteristics. Radiation characteristics describe how an antenna radiates or receives energy from the environment. These include radiation patterns, gain, directivity, polarization, and effective aperture. The behavior of the antenna by looking into its terminals is described by input characteristics such as input impedance, reflection coefficient, voltage standing wave ratio (VSWR).

Directivity (D) describes the ability of the antenna to radiate or absorb more power from a certain direction. It is defined as the ratio of maximum radiation intensity to the average radiation intensity [101]:

$$D = \frac{P(\theta,\emptyset)_{max}}{P(\theta,\emptyset)_{average}} \tag{6}$$

θ *and* $\emptyset$ are elevation and azimuth angles, respectively, as used in spherical coordinates. The isotropic antenna radiates equally in all directions. It has a directivity of 1. In practice, no antenna is completely isotropic and hence directivity is always greater than 1. Antenna gain is obtained by multiplying the efficiency of the antenna with its directivity. The choice of antenna gain is based on the knowledge of RF sources. If the source location is known, a high gain antenna is desirable. Otherwise, a lower gain antenna can be used to obtain energy from all possible directions as it will offer a wide field of view [104].

A qualitative analysis of antenna gain can be done by looking at the radiation pattern of the antenna. It is a pictorial representation of the strength of radiations received or transmitted by the antenna in a three-dimensional space. Three-dimensional radiation patterns can be cut into two perpendicular planes to give a two-dimensional view of the radiation pattern, called horizontal and vertical plane radiation patterns. For linearly polarized antennas, the orthogonal set of planes may be chosen in a special manner where one plane contains electric field vector while the other contains magnetic field vector, both being orthogonal to each other as shown in Figure 5.

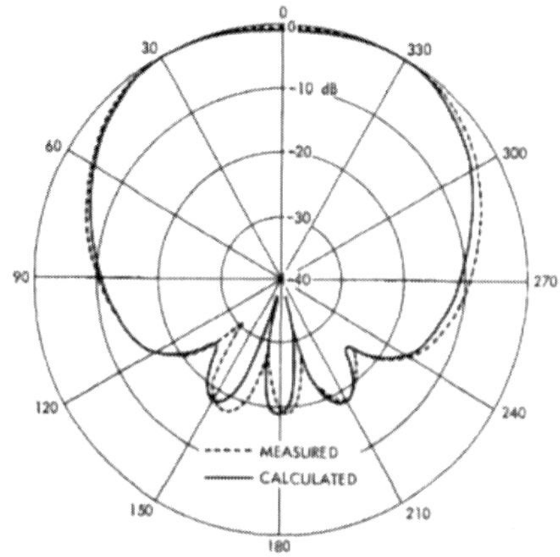

(a) E plane

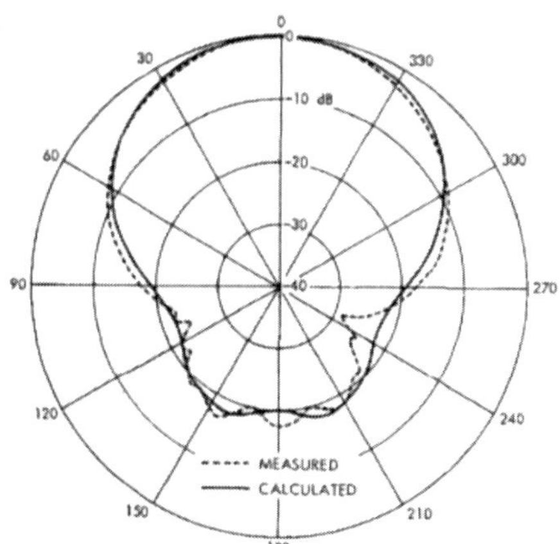

(b) H plane

Figure 5. E and H plane radiation pattern for a microstrip antenna [13].

Polarization is another important antenna parameter. Polarization describes variation in the magnitude and direction of the electric field of the radiated EM wave with space and time. Polarization can be broadly classified as linear, circular, or elliptical. If the polarization of the incoming signal and the antenna are completely matched, all of the input power gets absorbed. In case of polarization mismatch, only that part of radiations gets absorbed which has a component in the direction of the antenna's polarization. To achieve more reception of the signal, the antenna must be polarized the same as the polarization of the incoming RF signals. This reduces the reception efficiency. There is no reception of signal if the incoming wave and the antenna are orthogonally polarized. Circularly polarized antennas are widely used for reception as they can extract the power from at least some of the incoming wave's field components. Visual representation of polarization can be done using Poincaré's sphere [105]. Every point on the sphere represents a unique state of polarization. This method is very helpful for conversion between electrical and wave parameters of polarization.

For efficient power transfer from the antenna to the transmission line input impedance of the antenna must be matched to the characteristic impedance of the transmission line. Otherwise, the RF signal gets reflected from the antenna. The ratio of the reflected and the input RF wave is defined as Reflection coefficient, ρ. VSWR and return loss can be derived from the reflection coefficient as [101]:

$$VSWR = \frac{(1+|\rho|)}{(1-|\rho|)} \tag{7}$$

$$Return\ Loss = -20 \log |\rho| \tag{8}$$

Both these parameters characterize the reflection back of power. However, VSWR is usually more convenient to visualize. It represents the ratio of highest and lowest voltage points of the standing wave produced on the transmission line due to the superposition of incident and reflected waves. At the resonant frequency, the return loss reaches the highest (positive) value and VSWR attains a minimum value. A VSWR lower than 2 indicates a good match between antenna and transmission line at the described frequency. A 50 ohms transmission line is standard.

An antenna has a resonant frequency at which it receives maximally. At the resonant frequency, the impedance becomes minimum and purely resistive. Below resonant frequency, the antenna shows capacitive behavior and above this frequency, it shows inductive behavior. The resonant frequency is related to the physical size of an antenna. In general, the larger is the antenna lower is the resonant frequency. However, antenna size is constrained by the field of application. Body-mounted RF harvesting systems cannot have large-sized antennas. Multiband or wide bandwidth antennas are now in practice as the environment is abundant with more frequencies [14, 145].

Based on the above discussion a few considerations can be pointed out for designing an antenna in an RF harvesting system:

- Complete information on the frequency bands which are available in the working environments should be collected. This includes the variations in frequencies over time and space. Further design must proceed after this step.
- The strength of different frequency bands must be analyzed, again, over time and space. The antenna can be made to resonate at more than one frequency if more than one frequency has considerably high power levels, imparting multiband capability. However, for clear power maxima at a single frequency band of frequency, a narrowband antenna can lead to higher power absorption as compared to a multiband one.
- Antenna Gain must be high for RF sources that are either dedicated or which have the highest transmitted power in a particular direction. Otherwise, a more isotropic antenna can be used for randomly distributed sources as in these works [34, 35].
- Almost every time a circularly polarized antenna will give the highest received power in a random environment. However, if using a dedicated source, antenna polarization can be easily matched with the corresponding source's polarization for better reception.

Various types of antennas are available, patch antenna is the most common antenna used in RFEH systems because of their easy fabrication and implementation on the circuit boards. RFID chips also use patch antennas because of their miniaturized forms. They have a dielectric sheet with a metallic patch at one side and a ground plate at the other side.

Circular and rectangular patches are conventional shapes used; however, the patch can be fabricated in any continuous shape. Figure 6 shows a typical patch antenna schematic [142].

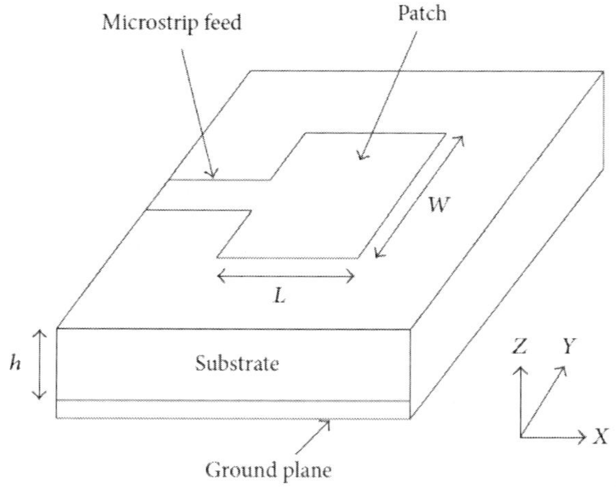

Figure 6. Patch antenna [142].

The performance of the antenna is greatly affected by the dielectric constant of substrate, height, width, and length of the antenna. The bandwidth of the patch antenna is directly proportional to the substrate thickness, h. For a length L and dielectric constant∈, resonant frequency for a patch antenna can be approximately given as

$$f = \frac{c}{2L\sqrt{\epsilon}} \tag{9}$$

where c is the speed of light.

3.2. Developments in Design

Many of the RFEH systems are deployed in compact designs; hence antenna miniaturization becomes an important aspect for RFEH systems. It may help make the system to be compatible with space constraints for the deployment of the harvester. Many methods have been proposed to reduce the size of a patch antenna. The simplest of these methods is to increase the substrate dielectric constant. Making slots in the patch is another commonly used method for achieving the same. Many slotted antennas have been proposed in literature as shown in Figure 7a [15], b [16], c [17], d [18], e [19], f [106], g [119]. In [15] two unbalanced circular slots were made on the diagonal of a circular patch antenna. The slots were made on the diameter making 45 degrees in the clockwise direction. As a result, the antenna radius was reduced from a theoretical 16.5mm to the proposed 15.5 mm, a decrease of 12 percent in size.

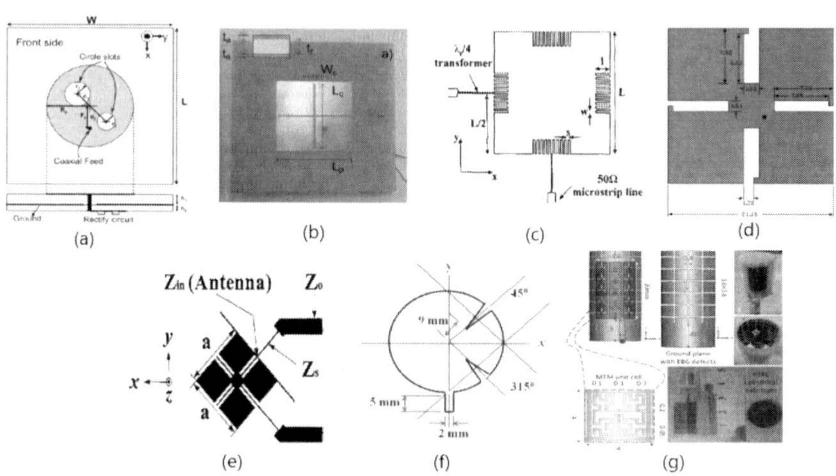

Figure 7. Miniaturized patch antenna designs [15, 16, 17, 18, 19, 106, 119].

Patch antennas suffer from low bandwidth and low gain. It is observed that high gain antennas give better conversion efficiency. Hence multiple antennas can be integrated to increase the gain of the receiver. This setup of antennas is called an antenna array. This setup allows the enhancement of the radiation and reception characteristics of the system. Various antenna arrays have been proposed in the literature. They contain antennas of different shapes and in different arrangements designed to meet specific requirements. Figure 8 shows a 20 element linearly polarized antenna array [20]. It was designed to scale the antenna gain and frequency and it could work in frequency bands of 915MHz and 2.45GHz. In [21] (Figure 9) a beamwidth enhanced antenna array was presented. Two additional auxiliary antennas were combined with a 1×4 antenna array. Beamwidth and PCE of the array showed a significant enhancement as compared to a single antenna element and a simple 1×4 antenna array. More than 50% efficiency was achieved for an incident angle from -38 deg to 35 deg.

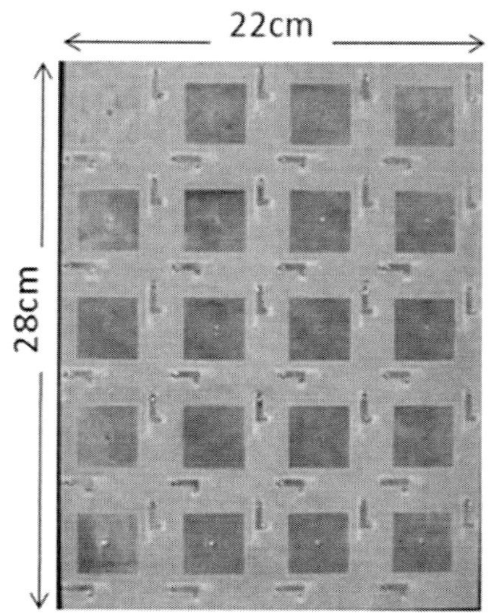

Figure 8. A 20 element antenna array [20].

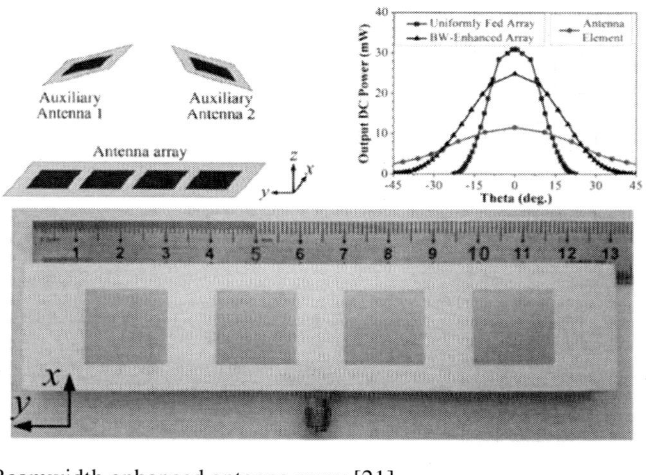

Figure 9. Beamwidth enhanced antenna array [21].

However, an antenna array is not feasible for applications requiring a small system size. Also in some applications, the rigid structure of the patch antenna may be undesirable, especially if the antenna has to be used in wearable systems such as health monitoring. A flexible patch antenna has been designed to meet such requirements. Such an antenna was designed in [22] with a triangular patch by using inkjet printing. PET was used as flexible material for substrate. It was a dual-band antenna that operated in regions of 15.9–19.7 GHz and 22.5–34 GHz as shown in Figure 10. Parylene is another material for making flexible substrates as used in [23]. A wrist band RF harvester was designed in [24] by using a flexible fabric antenna as shown in Figure 11. Various types of fabrics such as Kapton, Polyester felt, woven polyester, Cotton, etc were tested and analyzed at a frequency of 2.45 GHz. The extensive research shows a methodology for the development of flexible wearables for the practical world. The system could work with a considerably low power level of –24.3 dB and has a maximum end-to-end efficiency of 28.7% at −7 dBm. Such researches are paving way for making wearables a widespread reality. On similar application grounds, a wearable T-Shirt was designed in [121] by using 16 and 81 element rectenna arrays for harvesting power densities of 4–130-μW/cm2 and frequencies from 2 to 5 GHz, achieving up to 32% efficiency. The simulation also quantifies the effect of washing, body curvatures, and air between body and T-Shirt on the harvester performance. The proposed system diagram is shown in Figure 12. However, wearable antennas need to be more robust to physical forces, maintaining the acceptable radiation and reception characteristics despite the action of bending, compression, tension, and shear acting on it. They may also need to be robust to particular conditions they are subjected to, such as the washing of a wearable T-shirt might cause deterioration in its properties. The efficiency of wearable antennas is usually considerably lower than those of conventional patch antennas. But the research in wearable technologies is constantly trying to bridge this gap.

Figure 10. PET based patch antenna [22].

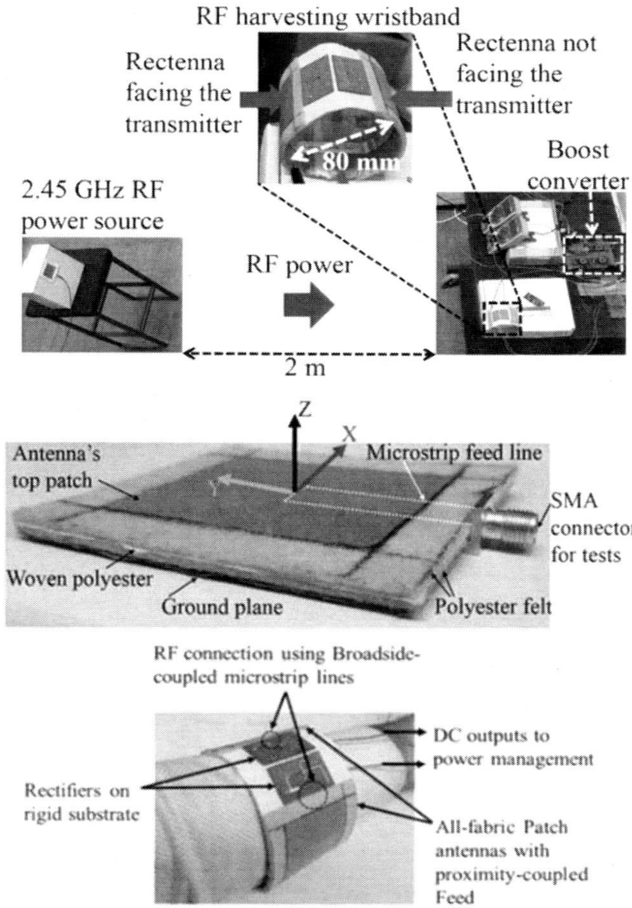

Figure 11. Fabric based patch antenna used in wrist bandtype of harvesting system [24].

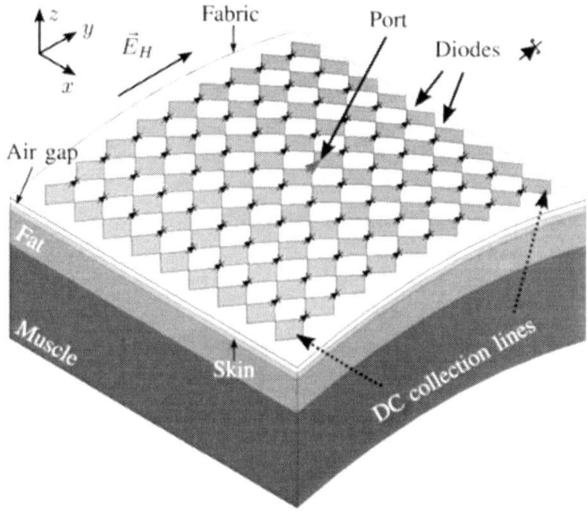

Figure 12. Rectenna (Rectifier+Antenna) array on a wearable T-Shirt.

Recently, metamaterials-based antennas have gained much attention as miniaturized RF energy collectors. These materials have unique properties like negative permittivity and negative permeability. They can provide higher efficiencies, wide operating ranges, and smaller sizes as compared to conventional patch antennas and antenna arrays. Hence they can be an excellent choice for a multitude of applications like IoT, WSNs, and RFIDs. Metamaterials have been reviewed and various designs have been explored in detail in this work [122]. Because of these merits, further researches in these materials can be a great boom for the future of RF harvesting, especially for wearables.

4. IMPEDANCE MATCHING

The RF energy is converted into electric energy by an antenna. This energy is then fed to a rectifier for converting AC signal to DC (or pulsating DC) signal. But if there is an impedance mismatch between the output terminal of the antenna and the input terminal of the rectifier, then some part of the energy captured by the antenna gets reflected from the junction and hence complete energy is not transmitted to the rectifier stage. This reduces the efficiency of the power transfer for this stage and hence for the complete harvesting system as well. This reflection also gives rise to standing waves between the antenna and the rectifier [25] causing a reduction in output voltage. Therefore, the standard 50ohms source (antenna) has to be matched to the input impedance of the rectifier [26]. Hence impedance matching is required between these stages.

An antenna can be modeled as a voltage source Vs in series with an impedance of value Z_{ant}. The load to the antenna (i.e., combination of further stages) can be denoted by impedance Z_{in}as shown in Figure 13 [27]. From the maximum power transfer theorem, the power delivered to the load will be maximum if the Z_{in} is conjugate of $Z_{ant}$and hence there will be no reflections. To fulfill this condition, we deploy a matching circuit before the rectifier. The circuit is chosen such that the combined impedance of the matching circuit and further stages sums up to Z_{ant}conjugate.

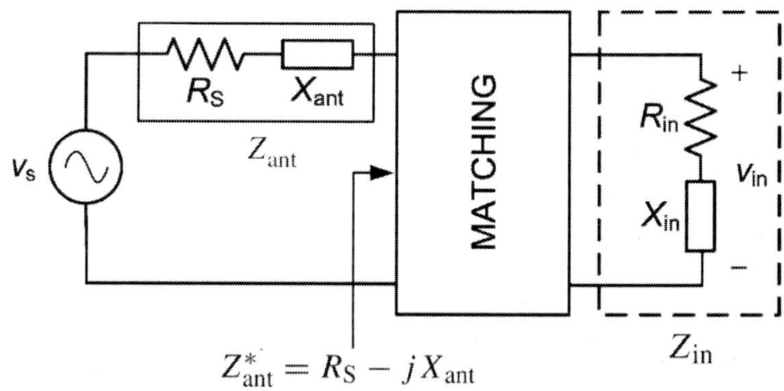

Figure 13. Matching circuit [27].

These matching circuits can be made either using a transmission line or lumped elements. Lumped elements include resistors, capacitors, and inductors. In theory, though, any arrangement of blind elements or transmission lines can act as a matching circuit [28]. However, L, T, and π networks are the most common topologies for impedance matching as shown in Figure 14.

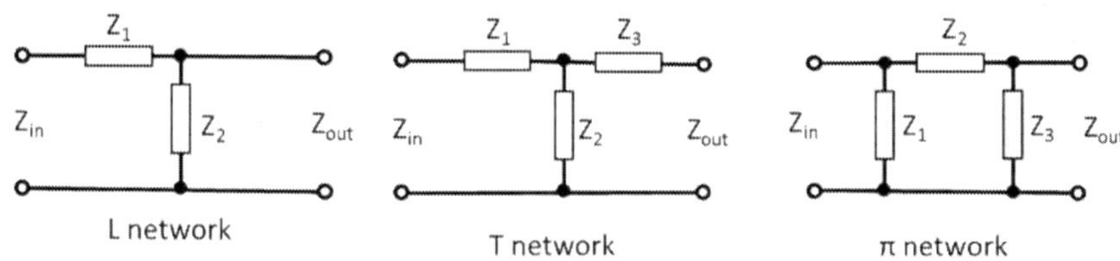

Figure 14. Matching network topologies [12].

L network consists of a series inductor with a shunt capacitor or vice versa. L section provides passive voltage amplification and undergoes lower losses as compared to topologies involving a higher number of components [29]. The bandwidth of the π network is wider as compared to the L network but the efficiency of the L network is more than that of the π network [30]. Because of an additional component in the π network, the Q factor can also be optimized in addition to impedance matching [25]. For the π and T-structures, the tenability is equivalent at the different frequencies [31]. Moreover, the T network can improve output voltage levels [12]. Experiments carried out in [26] clearly show that by adding L, T, and π matching circuits the efficiency of the harvesting system increases significantly. This is shown in Figure 15. For a 2 level Dickson voltage multiplier, L and T topologies show better performance for an input power range of around -15 to 10 dBm. However, for a 6 level multiplier, all the three topologies perform almost equally well in terms of efficiency.

For matching resistive components of impedance, a single resistor can also be used. But resistors lead to inevitable power loss [30]. Transformer matching [40, 41] is another impedance matching topology. A transformer converts power from one voltage and current level to another voltage and current level. As compared to LC matching, transformer matching requires a lower die area, since the transformer primary and secondary can be stacked [98]. Also, it is more robust to changes in characteristics due to parameter variations [98]. However, if the load has a high reactive component, a transformer would not be able to match it [30]. In addition to that, a transformer is not much effective at RF frequencies [30] and is not suitable for RFID chips because of its cost. Simple shunt inductors or capacitors can also be used for impedance matching. Generally, RFID tags use shunt inductor and L network topology because they are simpler and cost-effective as compared to other techniques.

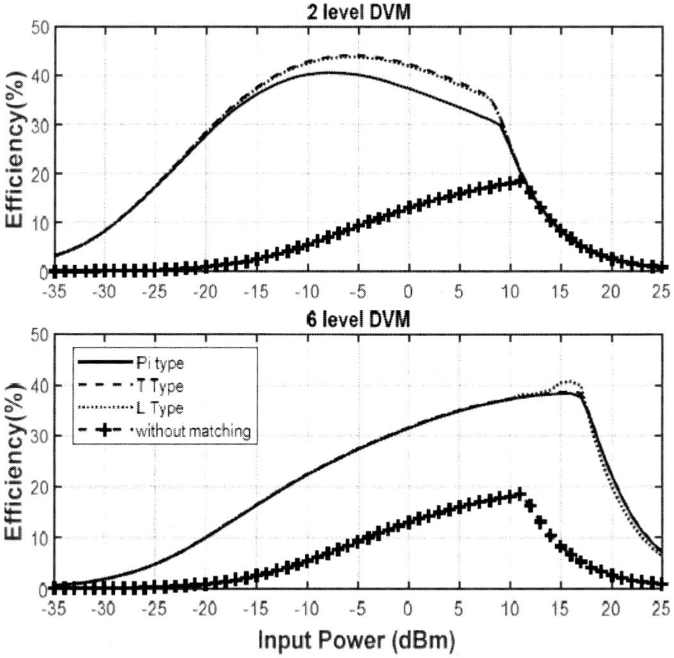

Figure 15. Efficiency of L, T, and π networks [26].

At high frequencies, lumped elements have parasitic reactance. For example, an inductor may behave like a capacitor if operated above its resonant frequency. That is why realizing reactive elements are difficult at high frequencies. Transmission lines perform well at higher ranges. Hence, the design of lumped elements can be done by using traces of the transmission line. Short circuited and open-circuited stub lines can be easily used as a substitute for reactive elements. For a phase constant of $\beta = 2\pi/\lambda$and characteristic impedance of transmission line as Z_o, Figure 16 shows the realization of reactive elements using transmission lines.

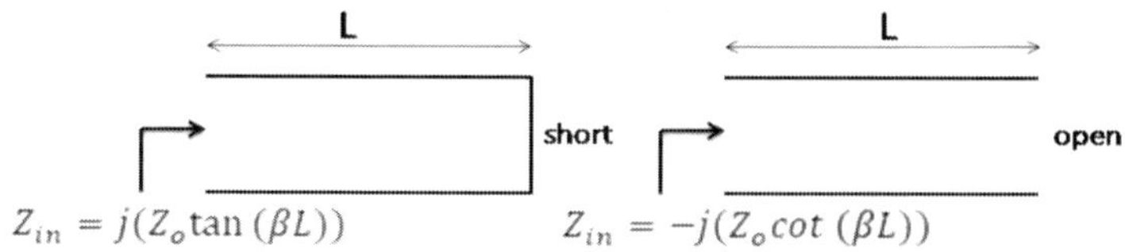

Figure 16. Complex impedance realization using transmission line [102].

Transmission lines may be an attractive option for impedance matching because of their low costs. When the transmission line is used, we deal with distributed circuit elements measured in units of value per unit length. The most widely used transmission line for this purpose is the microstrip line. Quarter wavelength transformer and stub matching techniques are some of the commonly used techniques for impedance matching using transmission lines. These methods exploit the smith chart for graphical calculation of the length of transmission line required for matching. The quarter wavelength method can match only resistive loads as shown in Figure 17. In this method, a piece of a different transmission line of length $\lambda/4$is connected before the load. Assuming lossless line, characteristic impedance, Z'_o of this extra transmission line is given as$Z'_o = \sqrt{Z_o Z_L}$. This method requires a specific value of the characteristic impedance of the transmission line used for the transformer which may not be of a standard value. This is a major disadvantage of using this method. This method can be extended for matching complex loads by connecting some length of the original transmission line between the transformer and the complex load. This extra length can be calculated by using the smith chart.

Majorly, two kinds of stub matching techniques are used, single stub matching and double stub matching. In a single stub matching, as shown in Figure 18, length $l$is responsible for matching the real part whereas length d matches the reactive part of the load. Length $l$depends on whether the stub is open or short at the end, and can be determined by using a smith chart. A stub, whether short-circuited or open-circuited, always has a reactive impedance which cancels the reactive part of load impedance, leaving just the resistive part at stub-main line interface. For impedance matching, this resistive value must be equal to the characteristic impedance of the transmission line. This requirement is fulfilled by the line $l$and the impedance is matched. Unlike the quarter wavelength transformer, the stub can be designed using a standard 50Ω transmission line for both mainline and stub [32]. Figure 19 shows a single stub matching technique for connecting the antenna and the rectifier section [7]. However, this method has the drawback of need to change length $l$as the load changes. This setback is taken care of in double stub matching where the location of the stubs are fixed and only the length of stubs has to be adjusted for matching a varying load. Nonlinear transmission lines [33] have also been used in literature for impedance matching.

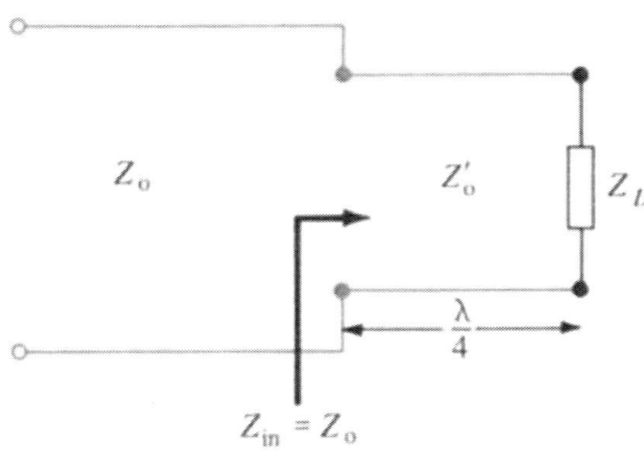

Figure 17. Quarter wavelength impedance matching [96].

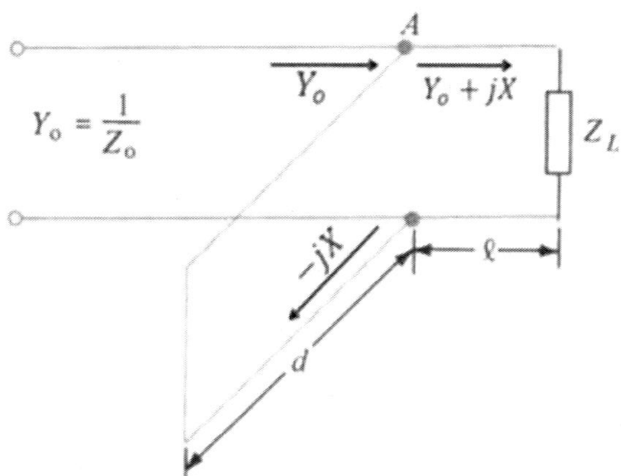

Figure 18. Single stub impedance matching [96].

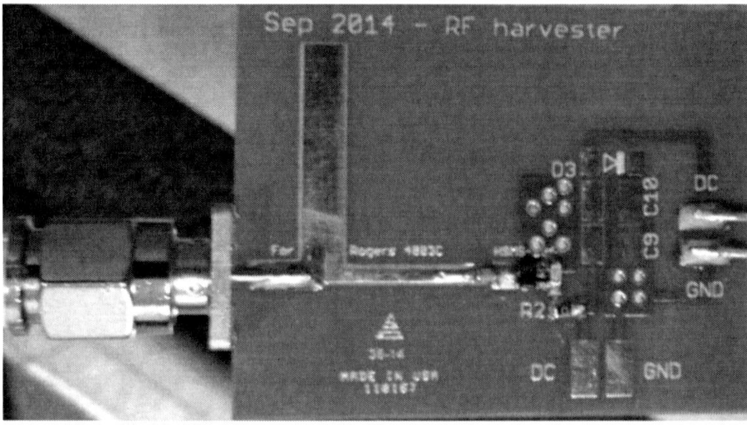

Figure 19. Single stub matching [7].

The design of Transmission line circuits is very closely dependent on the frequency of operation. And it is not possible to readily change the stub lengths to get a match with a new frequency. However, matching circuits using transmission lines can be used for tuning with multiband frequencies as shown in [32].

4.1. Q Factor and Tuning

The q factor is an important parameter for a Matching network. It is indicative of the selectivity of the circuit towards a particular frequency. Mathematically it can be given as [28]:

$$Q = \frac{Resonant frequency}{Bandwidth} \quad (10)$$

A high Q factor indicates that at the resonant frequency, the circuit swings with a higher amplitude as compared to other frequencies. It also indicates a lower power loss with respect to the power stored in the circuit. High Q is particularly important when the input power is concentrated in a narrow frequency band and the power level is low. Its value is generally limited by the individual components, especially the inductor which has a maximum Q factor of just 200 [28]. This causes damping of the input power. Authors in [28, 34] designed a high Q factor matching circuit by exploiting the fact that a crystal resonator can be used as an inductor in the region between its series and parallel resonance frequency. A capacitor and resonator were arranged in L-section topology. The Inductive region in crystal resonators can achieve high unloaded Q values of up to 80000. A comparison of high Q and low Q matching circuits as done in [28] is summarized in table 4. However, the disadvantage of a very high Q circuit is that the circuit becomes very sensitive to impedance variations.

Table 4. Comparison of a high Q and low Q matching circuit [28]

Parameter	High Q Matching circuit	Low Q Matching circuit
Voltage gain	49.5 at resonant frequency	21(constant)
Output voltage	3.6 V at resonant frequency	300 mV(constant)
Efficiency with 1M resistive load	30.3%	13%
Efficiency with 100k resistive load	39.74%	33.85%

The selection of a matching circuit depends on the input and output impedances to be matched. The problem is simplified if the impedances to be matched are constant. But RF harvesting has its own challenge. The impedance of the rectifier stage is a function of RF power, frequency, and load current. Therefore, a matching circuit needs to be designed by keeping these specifications in mind. The input power can vary over a wide range in a practical working environment. Also, due to the use of diodes in rectifier circuits, its impedance is highly non-linear; the voltage drop across the diode is a function of current if it operates under low power. These factors pose a great challenge in the designing of an optimum matching circuit. In [35] authors derived a mathematical formula for calculating the input impedance of the rectifier as a function of load current.

Impedance matching circuits can be either fixed or tunable over a range of input frequencies and power. Fixed matching circuits are designed for operating closely with a predetermined value of load conditions and Power level of the input signal. However, the input impedance of the rectifier stage varies greatly with the load and the input signal's frequency and power variations [36]. Figure 20 shows the variation of input resistance and capacitance of a rectifier with varying input signal power. At the input power level of about -33dBm, the input resistance is as high as 13.9 kΩ. The resistance decreases rapidly with increasing power levels. It is equal to 5.8 kΩ at −30 dBm, 620 Ω at − 22.6 dBm and goes down to 96 Ω at −9.6 dBm. As the input power varies from −30 dBm to −10 dBm, the capacitance varies from 0.4 to 0.56 pF. Hence if the working condition of the circuit changes from optimum values then the efficiency of these matching circuits degrades.

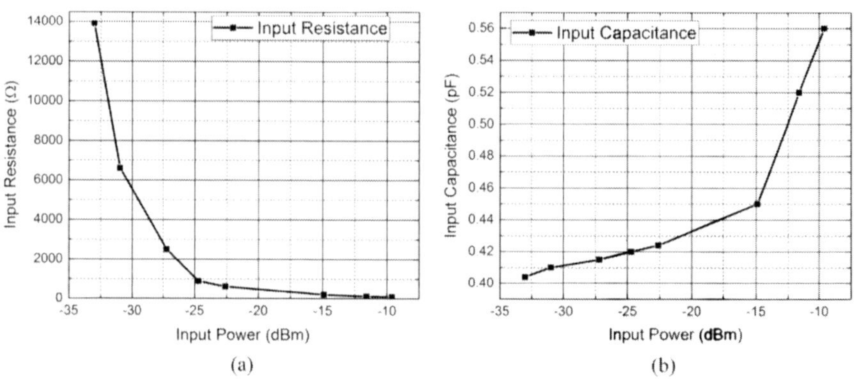

Figure 20. (a) Input resistance and (b) Capacitance variations of a rectifier with input power [37].

Many impedance matching circuits are designed for tuning with a narrow frequency band and fixed input power level. This type of design may perform well if the system is excited with a suitable input signal but their performance falls in real-life applications where the input signal varies widely. Thus for a practically efficient RFEH system, the matching circuit must be able to work in varying input conditions [38] shows a matching circuit design concept that can be tuned with a varying load of 1μA to 10μA. The circuit uses an L network with a variable inductor and capacitor. This kind of matching is particularly important for RFID chips which require different power while doing different tasks such as comparing, read/write and idle [5] conditions. In [39] a strategy to increase the power efficiency at variable input power is presented. The method required knowledge of the power density function of the RF signals. An on-chip rectifier was designed to be tested with the proposed matching circuit. The presented system maximized PCE over a wide range of input power levels with a maximum PCE of 32 percent at an input power of -15dBm.

The authors in [36] developed a feedback-controlled circuit that compensates for variation at the interface between the rectifier and the high-Q loop antenna by bringing the two at resonance. The control loop consists of a counter, comparator, and differentiator, and FSM blocks. The loop also helped in passively amplifying the antenna voltage and hence increased its sensitivity. Maximum end-to-end efficiency of 40% was achieved at a load of 0.33M Ω. Figure 21 shows the block diagram of the system.

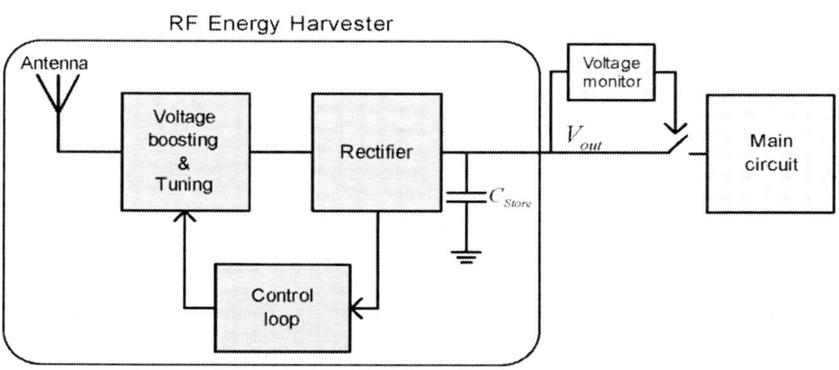

Figure 21. Feedback loop controlled matching [36].

One option to increase the acquired power is to increase the bandwidth of the matching circuit. However, from the Bode-fano limit, increasing the bandwidth of the matching circuit also increases the reflection coefficient of the rectifier stage. Adopting a different approach from this, a reconfigurable matching network was designed in [42] by using varactor diodes in a π network. The matching was done by tuning circuit components to get a match at the best available frequency. It was tested in a relatively narrow band of UHF TV broadcasting frequencies.

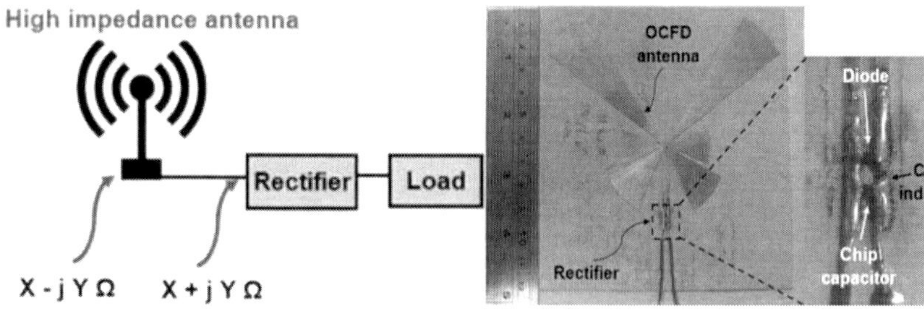

Figure 22. Impedance matching without a dedicated circuitry.

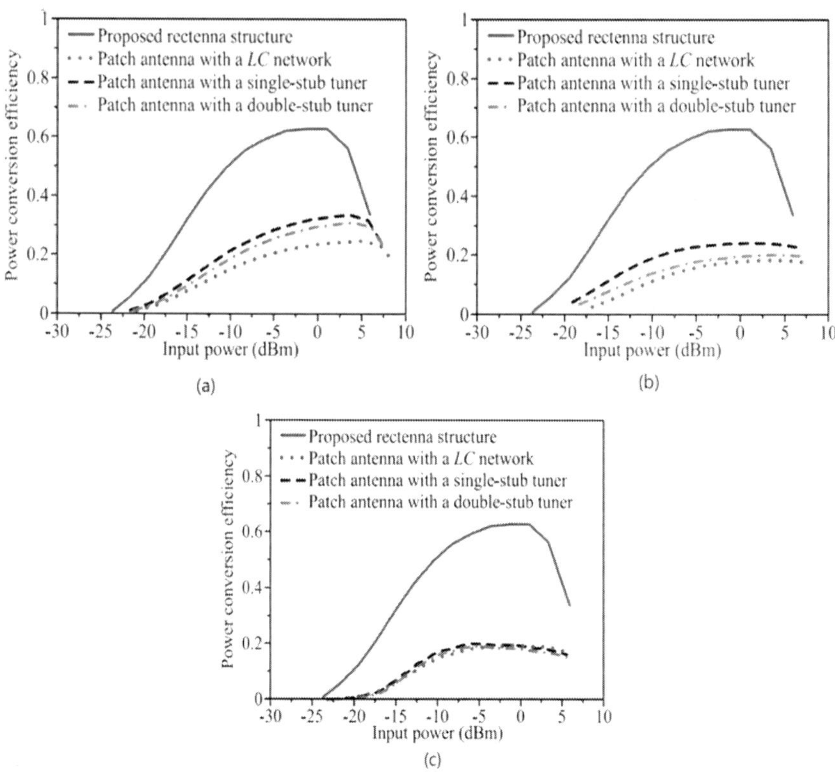

Figure 23. Comparison of various matching topologies with the proposed system in [136] for the rectifier used (a) Half-wave rectifier. (b) Single-stage voltage multiplier. (c) Dickson charge pump.

However, sometimes impedance matching circuit may not be required. This can be done by making antenna impedance to be conjugate of the corresponding rectifier's input impedance as done in [135], shown in

Figure 22. In [136] a design of the antenna-rectifier structure was proposed which removed the necessity of using a matching circuit. Several typical matching circuitries such as LC networks, single-stub tuners, and double-stub tuners as well as rectifying circuitries were tested and these were compared with the efficiency of the proposed design. The results have shown that for low power applications, using a matched antenna rectifier pair is capable of achieving higher efficiencies. This design methodology helps to eliminate the inevitable losses contributed by the matching circuit. This was further validated by the results which showed that the proposed design achieved efficiencies of 61.4%, 50.7%, and 31.8% at input power levels of -5 dBm, -10 dBm, and -15 dBm respectively at a frequency of 2.45 GHz. The variation in PCE as compared to other circuit topologies is shown in Figure 23. As analyzed in that work, these efficiency values are considerably higher than those in previous works.

Here are some points to pore over while selecting a matching circuit design:

- The selection and design of a matching circuit should be done keeping in mind the circuit characteristics and environmental characteristics like power levels.
- Analysis should be done on how prone is the circuit to variations in this power level, and accordingly, a compatible matching circuit must be designed. A study of variation in input impedance of a rectifier in the ambient environment can be carried out. This gives an understanding of the matching circuit which may be required
- If possible, non-resistive circuit elements should be used which do not dissipate much of the harvested energy as heat.
- If the variations are not too dramatic, then a simpler matching circuit with fixed calculated parameters can be used.
- If there are considerable fluctuations in the impedance value the efficiency will go down for a fixed parameter matching circuit design. In that case a reconfigurable or adjustable matching circuit may be used whose improved efficiency can compensate for the extra cost and efforts made in designing the matching circuit.

- For the standstill applications, the power level variations may not be a big issue as it is for wearables and other applications. Hence a tunable matching is very beneficial for the latter case.
- Low power applications may be designed without a dedicated matching circuit.

4.2. Reflection Coefficient and Efficiency

The basic purpose of the matching circuit is to prevent the reflection of input waves from the rectifier stage. The reflection coefficient quantifies this reflection of the signal wave from the rectifier input. It gives the ratio of reflected voltage to the input voltage and can be given by [96]:

$$T = \frac{Z_{rectifier} - Z_{antenna}}{Z_{rectifier} + Z_{antenna}} \tag{11}$$

where $Z_{rectifier}$ includes the impedance of the matching circuit. Hence, power reflection is proportional to T^2 and is given by [96]:

$$P_{reflected} = |T|^2 * P_{input} \tag{12}$$

Lower is the value of the reflection coefficient, lower will be the reflections from the rectifier. Therefore better is the matching and more is the efficiency of the system. The PCE of the matching circuit can be given as:

$$PCE = (1 - |T|^2) \tag{13}$$

5. Rectifier

The AC signal obtained from the antenna cannot be used directly for practical applications. Also, the incoming voltage level may not be sufficiently large to drive the load or charge a battery. So a rectifier is used to convert this AC signal into DC (or pulsating DC). In addition to AC-DC conversion, it may also boost the voltage level of the incoming signal to meet the requirement of a particular application. The most basic topology for AC rectification is a single diode rectifier, also known as the half-bridge rectifier as shown in Figure 24 [12]. When a positive AC cycle comes at the input port, theoretically the diode gets short and power is transmitted to the output port as such. In case a negative AC cycle comes, it gets blocked by the reverse-biased diode. Hence only positive half of the incoming signal can pass through, producing a half-wave rectified signal. Although practically, the peak of the output voltage is reduced by the turn-on voltage of the diode.

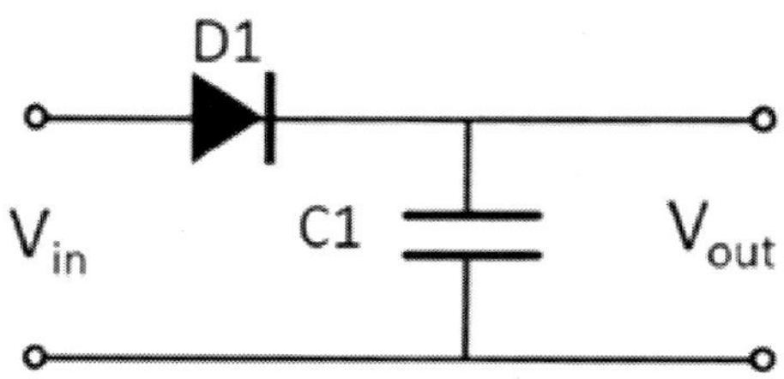

Figure 24. Half wave rectifier [12].

To obtain rectified negative half cycles in the output as well, a full bridge rectifier may be used. It is a four diode circuit as shown in Figure 25 [12]. When a positive AC cycle (i.e., point A is positive wrt point B) arrives at the input, diodes D2 and D3 get forward biased, whereas D1 and D4 get reverse biased. The current follows the path A-C-LOAD-D-B through the load. As soon as a negative AC cycle arrives, diodes D1 and D4 get forward biased, whereas D2 and D3 get reverse biased. The current follows the path B-C-LOAD-D-A. Current remains in the same direction for both the cases. Therefore both negative, as well as positive cycles of the signal, get rectified to produce a full-wave rectified voltage. Practically, the peak of the output is reduced by two times the diode turn ON voltage.

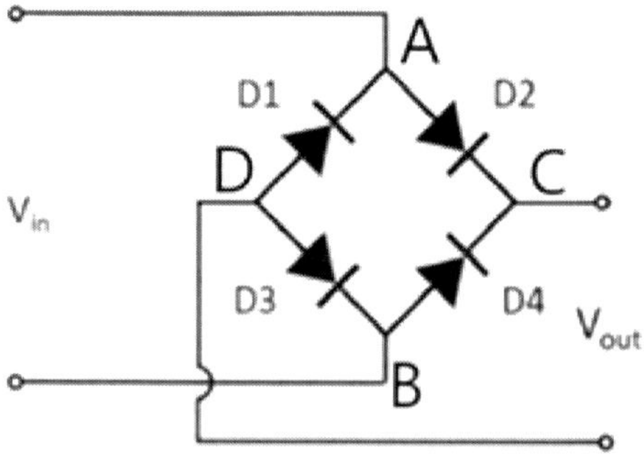

Figure 25. Bridge rectifier [12].

The peak output voltage in the above rectifier circuits is at max equal to the peak of input applied voltage. To increase the peak voltage of the signal special type of rectifier circuits are used. These rectifiers are known as Voltage Multipliers. They are typically made using capacitors. Inductor-based topologies are expensive and difficult to implement on IC [49], however, inductor-based boost converters have been implemented on PCB boards as done in [111] and shown in Figure 26. Cockcroft-Walton multiplier, also known as the Villard multiplier is one of the most popular topologies among multiplier circuits in RF harvesting. It can be seen as a series of several voltage doubler circuits. A single doubler circuit is shown in Figure 27(a) [44]. When a negative cycle arrives, Capacitor C1 charges to peak voltage Vp through diode D1. On the next positive cycle, D2 becomes forward biased and C2 charges through it. In this cycle, the voltage across C2 is the sum of source voltage plus the voltage, Vp across capacitor C1. Hence C2 charges to twice the peak of the source voltage. Several such stages of the voltage doubler can be stacked in series to increase the voltage further as shown in Figure 27 [b] [44]. A reconfigurable rectifier topology was designed in [120] using voltage doubler stages, shown in Figure 27 [c]. The design technique made it possible to boost the output voltage further by adding many such stages without much need to reconfigure the matching circuitry. An efficiency of 45% was achieved for the 900-MHz ISM band with input powers ranging from −15 to 20 dBm.

Figure 26. An inductor based boost converter [111].

Under no load, the output voltage is given by $V_{out} = 2n(V_{in} - V_{forward})$ where n is the number of doubler stages. In each doubler stage, there is a voltage drop across two diodes, hence the theoretical output of $2nV_{in}$gets reduced to $2n(V_{in} - V_{forward})$. As soon as a load is applied at the output, the voltage drop increases significantly from the predicted values. This is shown by the formula [103]

$$Voltage\ drop = \frac{I}{fC}(\frac{2}{3}n^3 + \frac{1}{2}n^2 - \frac{1}{6}n) \quad (14)$$

It is now a function of input frequency (f), load current (I), and capacitance (C) too. Also, the cubic dependence of a number of stages calls for using a reduced number of stages in the multiplier. The loading effect can be minimized by using capacitors with larger capacitance values. However larger capacitors will take a longer time to charge. If the stray capacitances at coupling nodes are significant fractions of stage capacitances the efficiency of conversion further deteriorates [46]. That is why it is not suitable for applying on monolithic integrated circuits [46].

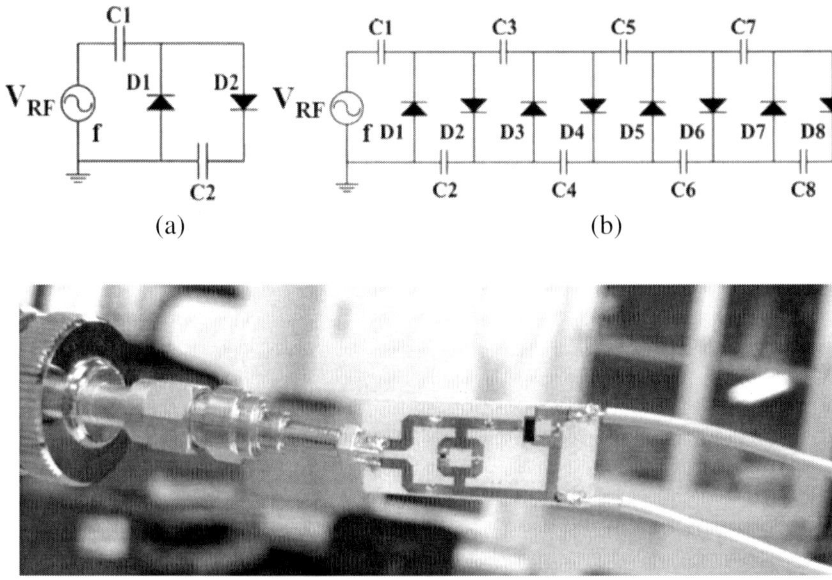

(a) (b) (c)

Figure 27. (a) Voltage doubler [44] (b) Multistage Cockcroft-Walton multiplier [44] (c) A reconfigurable voltage doubler [120].

Dickson multiplier, also known as Karthaus-Fischer cascade voltage multiplier, is another common voltage multiplier topology and is a modified version of the Cockcroft-Walton multiplier. Originally used as a DC-DC booster, it was later adopted for RF-DC boost conversion [25, 47, 48]. Figure 28 [47] shows a typical Dickson multiplier. It uses shunt capacitors and series diodes to create stages of RF-DC booster [47]. Shunted stages help to reduce the parasitic effects [51]. When the negative voltage cycle is incident at the input, all vertical diodes conduct, and corresponding capacitors in their way charge up [47]. As the voltage swings positive, horizontal diodes conduct and push the accumulated charge through capacitors in their way. The voltage increases with the number of cycles till it reaches the maximum theoretical value of Vout = 2n (Vin-Vfwd) where n is the number of stages [47] shows that at lower power levels more than 2 stages of the Dickson rectifier are rarely useful in increasing the efficiency of the rectifier.

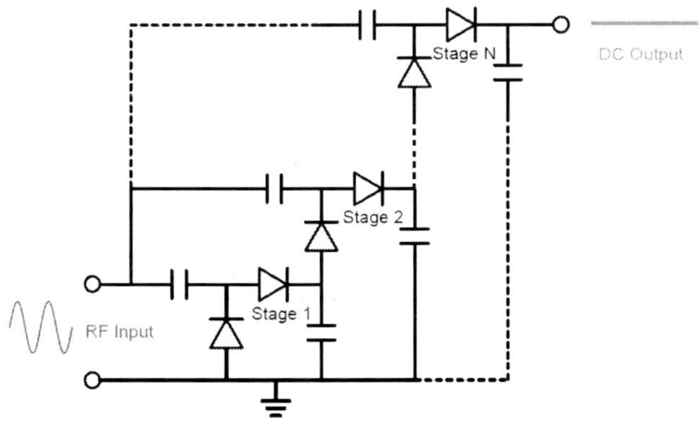

Figure 28. Dickson voltage multiplier [47].

Rectifier efficiency depends on several factors such as the microwave input power, the optimum connected load, the threshold voltage, and the breakdown voltage of the diode [52]. Variation of power conversion efficiency with input power is shown in Figure 29 [110], where Vj is the diode's forward voltage, Vbr is the diode's breakdown voltage and Rl is the load resistance. At lower power levels, the input voltage becomes comparable to the diode's turn ON voltage and hence output voltage falls. As the power increases, the efficiency of the rectifier increases and levels off with higher-order harmonics. As the input power levels further increase the voltage swings across diodes exceed the breakdown point. The breakdown effect is noticeable after the input power level of Vbr^2/4Rl and this results in a sharp decrease in the output efficiency [53].

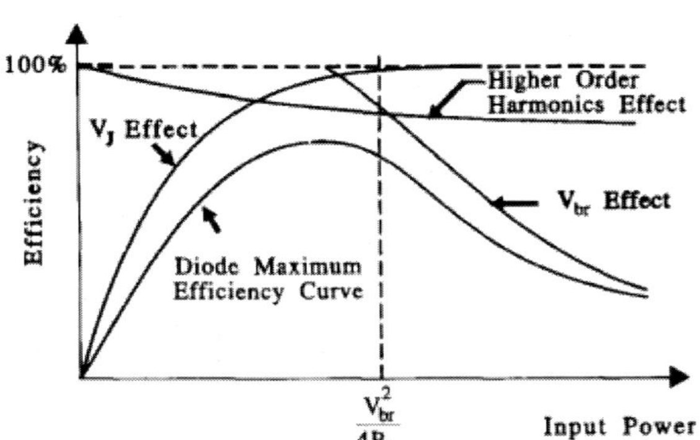

Figure 29. Microwave to DC power conversion efficiency v/s input power [110].

Input power will vary depending on the distance from the source. In a practical environment, this distance is not constant. Therefore the rectifiers should be efficient over a wider power range. An adaptive RF harvester was presented in Figure 30 [54] to extend the input power range for optimal Power conversion efficiency (PCE). PCE peaks were obtained at different power levels for single, parallel, and series arrangements of rectifiers as shown in Figure. The control circuit senses the input power level and switches the arrangement of rectifiers between single, series, and parallel to get the maximum efficiency possible.

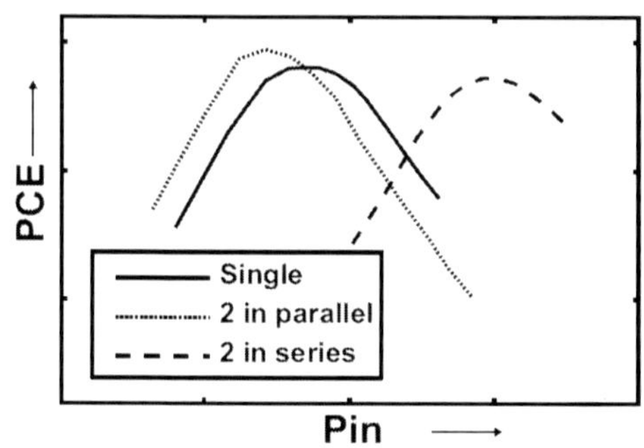

Figure 30. Efficiency v/s input power for an adaptive rectifier topology [54].

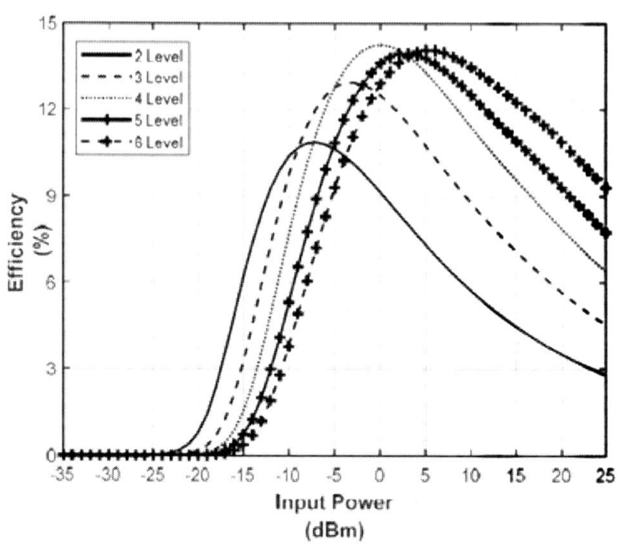

Figure 31. Efficiency v/s input power for different number of multiplier stages [26].

Efficiency is also a function of the number of stages of the multiplier. As the number of stages increases, maximum efficiency is obtained at higher and higher power levels. Figure 31 [26] shows this typical variation of efficiency with the number of stages for the Dickson voltage multiplier [26]. At a particular power level efficiency of the multiplier depends on the number of stages used in a multiplier topology. Hence an optimum number of stages can maximize the rectifier efficiency at a given power level. Due to the low power profile of RF signals, the number of stages in most publications does not exceed above 10 [51].

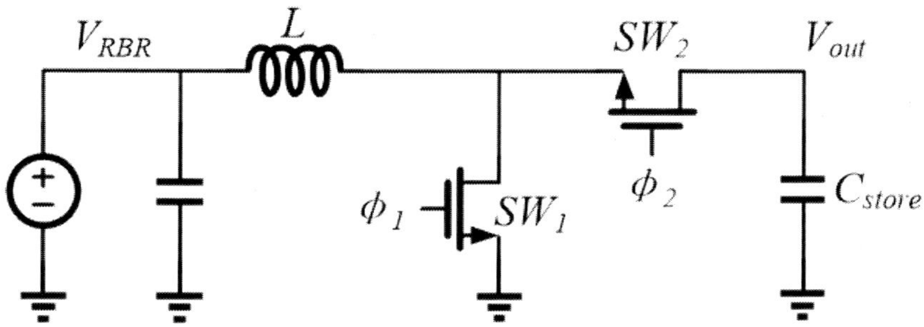

Figure 32. Boost converter for Impedance matching.

As diodes in a rectifier are non-linear devices, they produce harmonics of the input wave. These harmonics may be reflected by the antenna into the ambient space which in turn may be absorbed by the rectifier material [43]. This degrades the efficiency of the circuit. To block these harmonics as well as the DC generated from diodes, a bandpass filter may be deployed between the antenna and rectifier stages. This calls for the use of two matching circuits, one between antenna and filter and the other one between filter and matching circuit. However, this kind of circuit is not used in low power circuits such as RFID because of added cost and complexity. A low pass filter can also be placed after the rectifier to block any higher-order harmonics to reach the load [30, 43]. It can be made with a simple arrangement of Resistor and capacitor. Furthermore, matching can also be provided between rectifier and device. [134] designed a matching circuit in the form of a boost converter to deliver as much power as possible from the Rectifier to a 1-mF supercapacitor at a frequency of 2.45-GHz, shown in Figure 32. For a poorly matched load, such matching can further enhance the power conversion. However, these extra designs should not be used in low-power applications because of extra losses associated with the components added.

5.1. Diodes in Rectifier

RF harvesting system deals with high frequency and low power signals. This demands fast and low loss rectifier design. Diodes are the main components of a rectifier. They are the major cause of harmonics and power loss [55]. The characteristics of a rectifier depend largely on Diode characteristics. A diode with a large threshold voltage will reduce the output voltage level and hence the efficiency of the rectifier. Therefore diodes need to have a lower threshold voltage for better efficiency of the rectifier. Further, to cope up with high-frequency signals, diodes with high switching speeds are desired [30]. Schottky diodes are ideal to be used in RF circuits. They have threshold voltages ranging from 0.2V to 0.5V, lower than that of typical PN junction diodes. Because of the low junction capacitance, they have fast switching speeds. Additionally, Schottky diodes have low series resistance (Rs) and high saturation current [55]. For example, the SMS 7630 Schottky diode has a built-in voltage of 0.34 V and is suitable for use at frequencies up to 26 GHz and higher. But they have low breakdown voltages. Hence appropriate voltage levels must be applied to prevent them from breaking down. By optimizing the efficiency of Schottky diodes, ultra low power sensing and conversion performance were achieved in [56] for powers as low as -50dBm.

A Schottky diode is a metal-semiconductor junction diode, where metal acts as an anode and the semiconductor as a cathode. The equivalent circuit of the Schottky diode has two kinds of parasitic components, the chip parasitic and the package parasitic [57]. The package parasitics consist of an inductance and capacitive impedance [57]. Capacitive chip parasitic is the junction capacitance and resistive chip parasitic is the junction resistance. The diode also has a parasitic series resistance which represents the combined resistance of the bulk layer of the silicon substrate, bond-wire, lead-frame, etc. [12]. This parasitic series resistance generates losses in the diode. All parasitic components in the rectifier must be kept low to maintain a high overall system Q and thus power conversion efficiency [58]. A good diode model helps to simulate the rectifier with accurate results. In [55] authors find small-signal S parameters for Schottky diodes and characterized large signal diode equivalent circuit using numerical approximation method. Some off-the-shelf Schottky diodes were compared for their efficiency for the Dickson multiplier [47]. Results show that despite increasing series resistance, efficiency increases at low power levels with increasing saturation current. These same diodes perform poorly at higher power levels due to a lower breakdown voltage as shown in Figure 33 and Table 5 [47].

Table 5. Comparison of off-the-shelf discrete Schottky diodes [47]

Part Number	Manufacturer	Saturation Current *(microA)*	Series Resistance (ohms)	Breakdown Voltage *(V)*	Zero-Bias Junction Capacitance *(pF)*	Ideality Factor
SMS7630	Skyworks	5	20	2	0.14	1.05
HSMS2850	Avago	3	25	3.8	0.18	1.06
HSMS2862	Avago	0.050	6	7	0.18	1.08
HSMS2820	Avago	0.022	6	15	0.7	1.08
IPSIOSB82	NXP	0.010	5.6	17	0.978	1.029
STPS0520Z	STMicroelectronics	5.6	0.133	1.7	118	0.999

Figure 34 demonstrates other competitors of Schottky diode for rectification of RF signals. It also shows the improvement in these diodes over years in terms of Current responsivity. Improvement in this parameter helps improve the efficiency of a diode. It also accounts for the poor efficiency of Schottky diodes at low power levels of 1uW. It can be seen that backward tunnel diodes, having the best current responsivity, are the most efficient choice for RF rectifiers in present times [115]. These developments pave way for the substitution of Schottky diodes with more sensitive and efficient diode types.

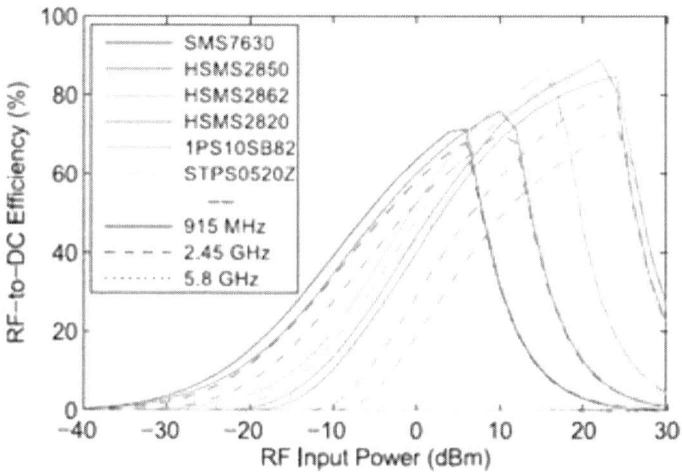

Figure 33. Comparison of different Schottky diodes for single stage Dickson multiplier [47].

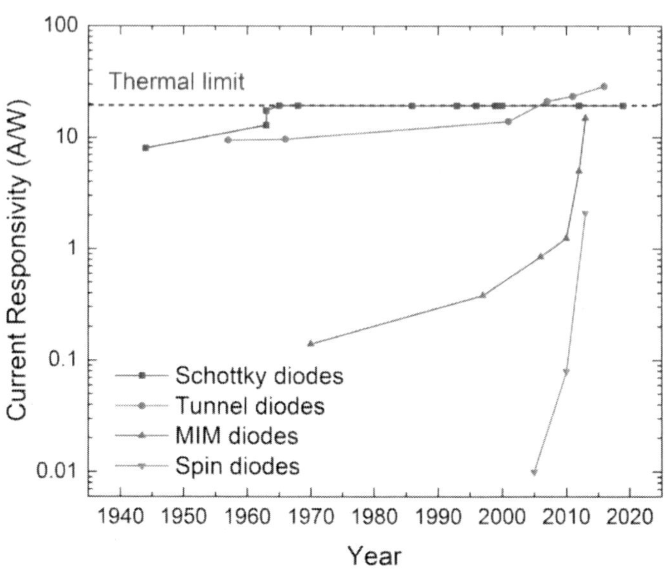

Figure 34. Improvement in Current sensitivity of rectifying diodes used for RF harvesting [115].

5.2. CMOS

High manufacturing cost is the major drawback of Schottky diodes [59]. That is why these diodes are not always available in CMOS technology as it increases the manufacturing cost considerably [59]. In addition to that, Schottky diodes have high zero-bias junction resistance and are not very efficient at microwatt power levels [51]. Developments in MOSFET technology have provided MOS transistors an alternative to Schottky diodes. For example, the Dickson multiplier can be integrated into an IC by using NMOS as shown in Figure 35 [51]. This design offers a low turn ON voltage and high power conversion efficiency. Because MOSFET technology allows custom designing capability and has more sensitivity to low voltages [51], many research groups have moved towards CMOS processes for the design of rectifier and voltage booster circuits.

Though CMOS may have lower threshold voltages then it also has high leakage current which reduces the efficiency of the rectifier. Hence there is a tradeoff between low threshold voltage and high leakage current [12]. Several works have been done to optimize CMOS-based rectifier performance. Various methods have been developed to reduce threshold voltage, Vth for MOS rectifiers [60, 61, 62, 63, 64] with a comparison of these methods in [64]. Reverse leakage current loss in the CMOS rectifier was improved in [65] by utilizing the novel reducing reverse loss (RRL) technique.

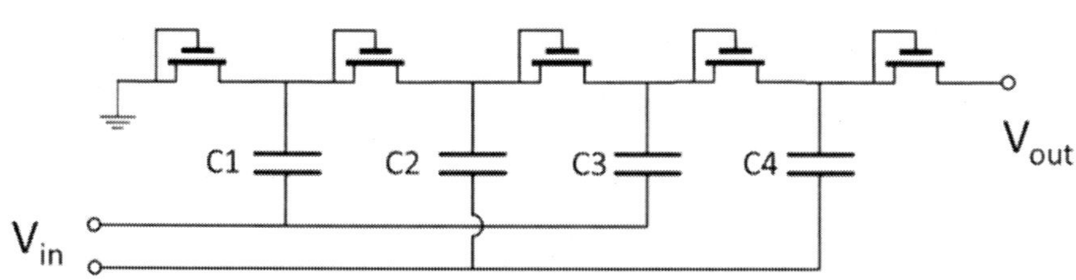

Figure 35. CMOS dickson charge pump [51].

[65] has cataloged several types of rectifier circuits which have been proposed in the literature to be integrated into standard CMOS processes. These are shown in Figure 36 [65]. Most of the designs suffer from the

high leakage current of the MOS devices. In addition to that, Vth-cancellation and active-diode topologies require an additional power source for their functioning. The fabrication size of the CMOS transistor is important for the efficient working of the CMOS rectifier. Larger is the size, higher is the parasitic capacitance. Smaller is the size, lesser current can flow through the transistor. Hence an optimum transistor size is needed for a tradeoff [12].

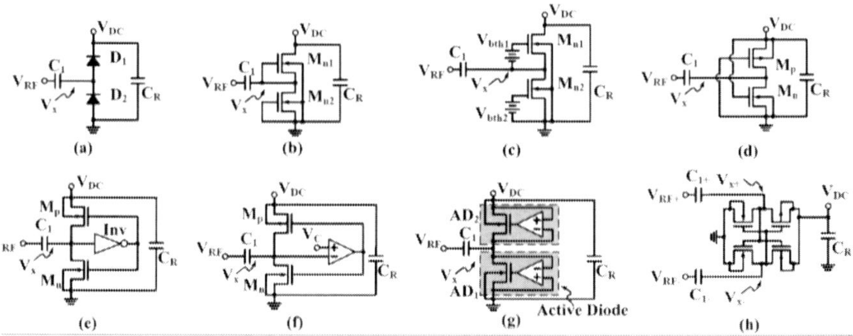

Figure 36. CMOS implementable rectifier topologies [65] (a) diode-type, (b) diode-connected MOS, (c) Vth-cancellation, (d) self-Vth-cancellation, (e) inverter-based, (f) comparator-based, (g) active-diode, (h) differential-drive.

5.3. Active Diodes

Active diodes as shown in Figure 37 [46] are good candidates to replace conventional diodes from voltage rectifiers [68, 69]. Using an active diode is a simple method for reducing the diode's forward voltage. An active diode is can be made by a combination of a comparator and a MOS transistor. If the voltage at Anode is greater than that at Cathode, the output of the comparator turns low. This turns ON the PMOS switch and the signal can pass through the Active Diode. In case Cathode is at higher potential with respect to Anode, the comparator output turns HIGH and hence turns the PMOS switch OFF. However additional energy supply is required to power the circuit. Hence the major challenge is designing a comparator that is power efficient and has low start-up voltage [65]. At higher power levels conduction loss become more dominant [46] and the efficiency of the diode further decreases. Several studies have been done to improve the efficiency of active diode [66, 67]. Various voltage multipliers topologies have been implemented using active diodes.

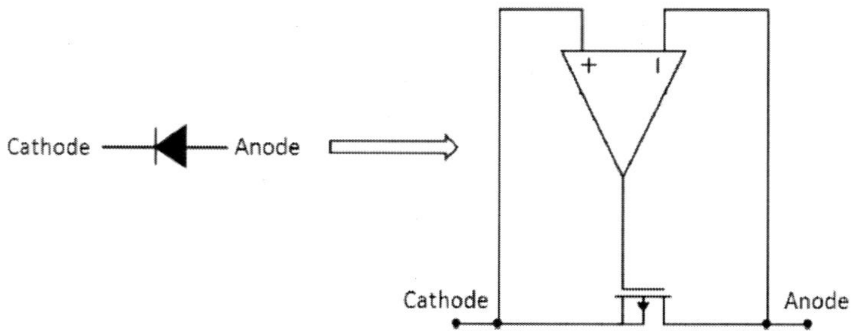

Figure 37. Active diode [46].

At sub-microwatt power levels, Schottky diodes and CMOS technology cannot perform well because the antenna cannot deliver enough voltage levels to reach the turn-on voltages. As a result, the rectifier goes into a dead state. To tackle ultra-low power levels we need even lower threshold voltages. Tunnel FETs (TFETs) have been proposed in recent times which seem to open the door for lower power harvesting (sub-0.25 V) [143]. These steep slope devices are potential candidates to replace CMOS from rectifiers [144]. In comparison to Silicon and Germanium devices, the design of TFET material with group III-V materials can reduce forward losses of the diode because of the ability to conduct more current with lower voltages.

The rectifier stage of the system is a key to achieve high PCE as most of the power is dissipated in diodes only. As we saw there are many parameters to be considered for designing an RF to DC voltage converter, it becomes quite challenging to address the need of the output voltage level required and maintaining good PCE at the same time. Hence here are a few things to be kept in mind:

- Rectifier losses are the deciding factor for most RF harvesting systems. Hence they should be minimized by trading off between parameters like threshold voltage, current responsivity, leakage current, etc. In addition to that it is known that using more stages in a rectifier leads to higher sensitivities and more voltage output but doing so also results in increased losses. Hence a good trade-off between sensitivity and PCE should also be made.

- Pick up the right Schottky diodes by analyzing the RF power variations at a location. The diode chosen should have its peak efficiency at the values near the ambient power level.
- CMOS should be considered when higher sensitivities are required and the design demands for custom designing. Unlike Schottky diodes, MIM diodes are now possible to be integrated with the CMOS process.
- When using CMOS technology for rectifiers a clever trade-off between threshold voltage and Reverse leakage current is of great importance.
- The design of the rectifier is dependent on RF power. At microwatt power levels Schottky diodes have high zero-bias junction resistance which degrades PCE. Hence one must look for other suitable diodes at such very low powers. Spin diodes may replace Schottky diodes for achieving lower threshold voltages whereas a replacement with Tunnel diode might result in increased efficiency.

6. Load/Energy Storage Devices

If Output Power from the rectifier section is high enough, it can be used directly to feed to a device. These are called autonomous systems [70]. Such a system was introduced in [71]. It was the first battery-free 19uW wireless bio-signal acquisition system powered solely from an RF power and/or thermoelectric harvester. But in case the power requirement of the load exceeds the continuous power delivered by the rectifier, it cannot feed the load directly. In that case, a storage unit is needed to first accumulate the power coming from the rectifier, and later give it to the output device. This is referred to as hybrid autonomous topology [70].

If RF energy is not enough to power the system fully, in that case, battery supplemented [70] topology can be employed. This system has a battery as the main source of power while RF energy is a secondary source of power. The battery used can be primary or secondary type. This type of system can be especially useful in areas that have a good amount of RF energy at one time and unacceptably low power at the other time. Figure 38 [70] shows typical block diagrams for the three harvesting topologies. The system complexity increases as we move from autonomous to hybrid autonomous to battery supplemented topology. Super-capacitors and Batteries are common energy storage devices.

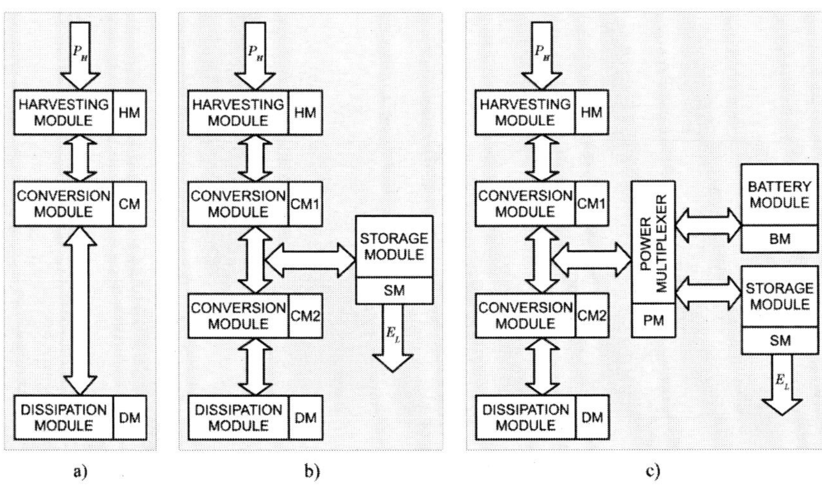

Figure 38. Topologies based on storage device used/not used [70] (a) autonomous harvesting system; (b) autonomous hybrid harvesting system; (c) battery-supplemented harvesting system.

A number of battery technologies with a range of storage capacity are available commercially. Common battery types include lead-acid, nickel-cadmium (NiCd), nickel-metal hydride (NiMH), and Lithium-ion (Li-ion). The selection of a proper battery depends on the application and should be done by considering various features of each battery storage technology. Energy density, internal resistance, depth of discharge, self-discharge, and tolerance to overcharging are important battery characteristics [70]. Lead acids are bulkier and hence less suited for RF harvesting. NiCd has low energy density but has a long lifetime and low cost. NiMH has more energy density than NiCd but this comes at the cost of a reduced lifetime. However, unlike NiCd, NiMH is non-toxic. Because of their lower cell voltages, NiCd and NiMH can be charged with lower voltage sources. Hence these batteries are desirable when the RF circuit can output low charging voltages.

Li-ion can provide high power density and high cycle life. However, High cost and proper charging requirements are major drawbacks of Li-ion. In case there is a deviation from the ideal operating conditions battery life of these batteries is reduces significantly. Furthermore, it can catch fire in case any damage occurs to the battery pack. On the other hand, NiMH batteries can be charged indefinitely using the trickle charge method [72] and hence requires less complex charging circuits as compared to Li-ions. Despite all its disadvantages, Li-ion is a popular and widely used battery type. New battery technologies are being introduced in the market continuously. Table 6 [70] compares some common battery technologies.

Table 6. Comparison of major battery types [70]

Type	Cell Voltage(V)	Capacity(Ah)	Specific Energy (Wh/kg)	Cyclic capacity
Lead Acid	2	1.3	30-50	500-1000
NiCd	1.2	1.1	50-60	10000-20000
NiMH	1.2	2.5	60-70	1000-20000
Li-ion	3.6	0.74	75-200	1000-100000

As the name suggests, supercapacitors are simply capacitors with very high storage capacitances (several Farads). They have power densities higher than that of a typical battery. However, their energy density is up to 50 times lower than typical batteries. In addition to this, they have a life of up to one million charge-discharge cycles, as compared to a typical value of just a few thousand charging cycles for typical batteries. Charge/discharge efficiency can be as high as 98% for supercapacitors [70]. However, Self-discharge is a major drawback of supercapacitors. Studies have shown that supercapacitors may have a self-discharge of up to 5.9% to 11% in a day [92, 93]. Table 7 [89] compares a typical supercapacitor with a Li-ion battery.

Table 7. Supercapacitor v/s Li-ion battery [89]

Function	Supercapacitor	Lithium-ion (general)
Charge time	1-10 seconds	10-60 minutes
Cycle life	1 million or 30,000h	500 and higher
Cell voltage	2.3 to 2.75V	3.6V nominal
Specific energy (Wh/kg)	5 (typical)	120-240
Specific power (W/kg)	Up to 10,000	1,000-3,000
Cost per kWh	$10,000 (typical)	$250—S1,000 (large system)
Service life (industrial)	10-15 years	5 to 10 years
Charge temperature	-40 to 65°C (-40 to 149°F)	0 to 45°C (32°to 113°F)
Discharge temperature	-40 to 65°C (-40 to 149°F)	-20 to 60°C (-4 to 140°F)

Recently though, a hybrid storage technology called supercapattery (supercapacitor+battery) has evolved [90, 91]. This technology combines the merits of both supercapacitors and batteries providing high energy as well as power density.

6.1. Battery Charging

Batteries are very sensitive to the charging process. Because of differences in chemical structure and cell voltage, each battery type requires a different charging process. A battery charging circuit is responsible for providing appropriate input voltage for the specific type of battery used. A novel trickle charging circuit was designed for the NiMH battery in [72]. After boosting voltage from about 200mV to 2V, a charge buildup was done by the charging circuit, and then it was sent to the battery in form of pulses. Each pulse is about 10msec in duration. The battery charger as shown in Figure 39. NiMH can be given these pulses indefinitely. Another NiMH charger was proposed in [7]. However, Li-ion batteries need a more sophisticated charging circuit than NiMH. [94]

proposed a charging algorithm for charging Li-ion batteries as shown in Figure 40.

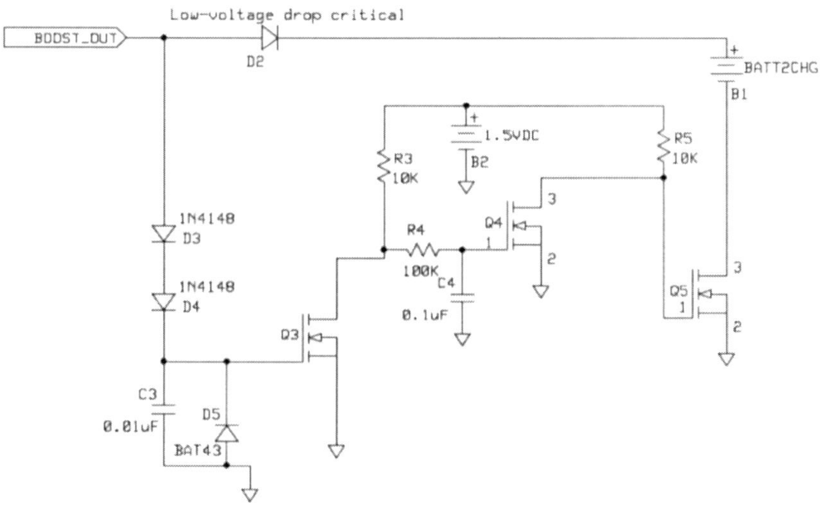

Figure 39. NiMH battery charger [72].

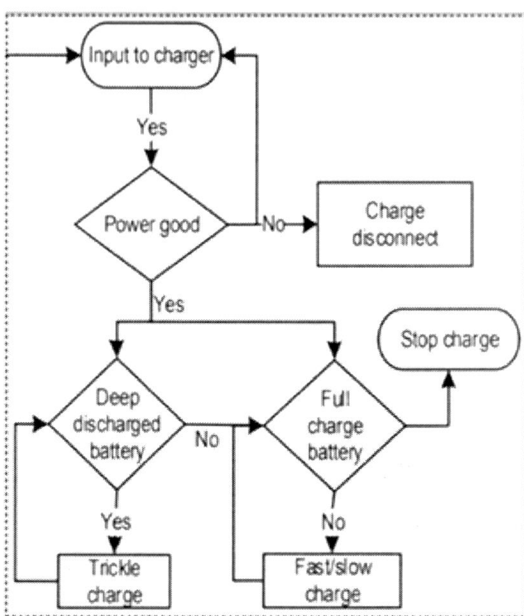

Figure 40. Li-ion battery charging algorithm [94].

6.2. Effect of Load on Efficiency

The magnitude of load impedance greatly impacts the overall efficiency of the system. Efficiency will deteriorate if the value of load impedance is too high or too low. Hence an optimum value of the load impedance has to be used to achieve greater efficiency. Various studies [25, 73, 26] have been done on the effect of load resistance on system efficiency. A comparison of various loads from 1k to 181k ohms was done in [73] for input powers ranging from -20dB to 20dB as shown in Figure 41. The results show that the system attains the highest efficiency when the load resistance is at an optimum value. For example, at about 10dBm both 10Kohm (lowest) and 160Kohm (highest) load values show lower efficiencies as compared to intermediate values of 60K and 110K ohm loads.

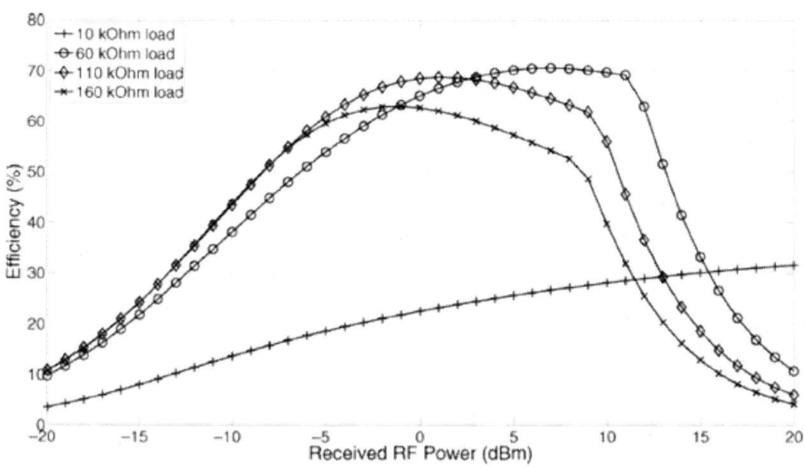

Figure 41. Efficiency of RFEH system with varying load [73].

The variations of efficiency due to varying load conditions can be suppressed to some extent by incorporating a resistance compression network (RCN) in the antenna-rectifier design [139, 140]. An RCN is composed of two parallel branches having equal variable loads in series with a reactive element, arranged such that the phase response of the upper and the lower branches are 180 degrees apart. Apart from restricting load variations, these networks can, in general, be used to reduce non-linearities in the circuit. A typical RCN network is shown in Figure 42(a) [141]. A novel impedance compression network (ICN), different from the RCN, was proposed in [114], shown in Figure 42(b). The authors have also shown the methodology to design the circuit. The design turned out to be simpler and more compact as compared to the conventional RCN. In contrast to RCN's ability to suppress only resistive loads, the proposed circuit was able to compress the complex impedance variations. Hence such networks can be of great help if non-linearities pervade the system.

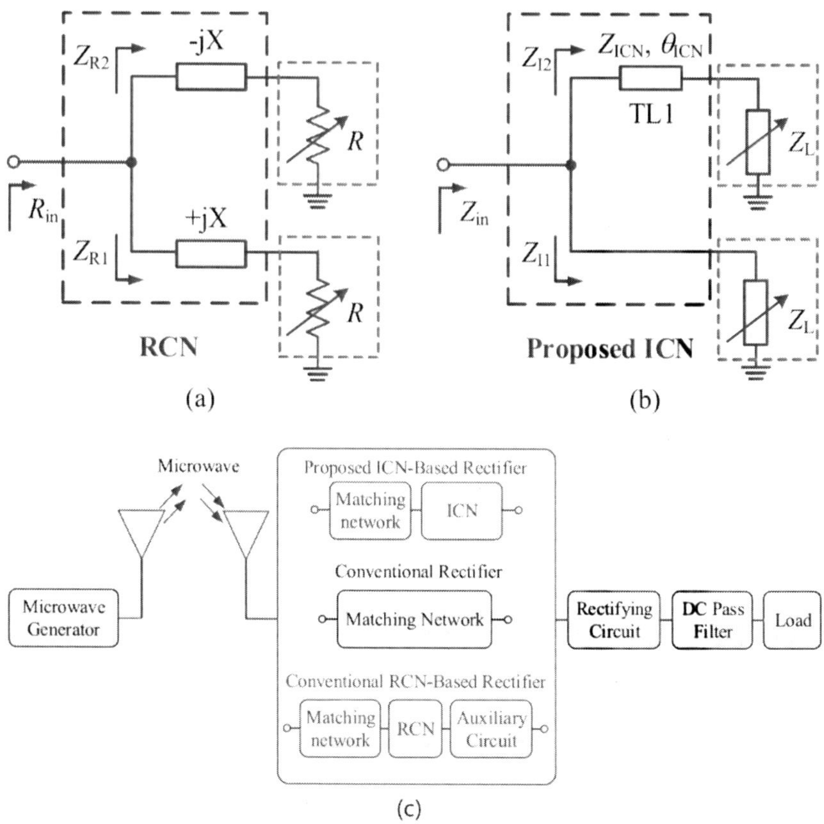

Figure 42. (a) Schematic of a typical RCN (b) Novel ICN proposed in [114] (c) RF harvesting system incorporating these systems.

Most of the time the whole system is designed by keeping in mind the load device and its power requirements in such a way that there is compatibility between the two. Many factors decide the choice of system elements based on load device and vice-versa. Here are a few considerations regarding the load device in an RF harvesting system.

- The maximum current level of the load device should be considered for it to be compatible with the system and it must be assured that there is no overloading.
- In case a storage element is used, the selection should be done considering the system's output voltage levels and cell voltage of the available storage elements.
- An analysis of the efficiency with different output load values should be done and, if possible, the load value closest to the optimum efficiency should be used.
- Storage elements, especially capacitors should be checked for leakage of charge. This effect can nullify the extra efforts made in increasing the efficiency of other parts of the system.
- Circuits for decreasing the effect of non-linearities can be used if cost and designing efforts can be compensated by the increased efficiency.
- Different storage elements have hazards of their own. Hence it should be assured that the environment in which the system is deployed should not provide an element of damage.

7. POWER MANAGEMENT

For efficient working of the overall RFEH system, additional circuitry may be used in the RF harvesting system. This control circuit may take care of a variety of purposes including charging of storage elements, controlling the flow of power, and switching between the sources (in case of hybrid autonomous or battery supplemented systems). The umbrella term for the circuit performing these functions is power management Unit

(PMU). A PMU may have different functions and organizations. It may help in supplying constant power to the load despite varying input RF power and changing load conditions [74]. They may also provide monitoring of system performance. Battery charging circuits can also be seen as a subset of a PMU.

At some time of the day input, RF signals may have insufficient low power to feed a device or charge a battery. Such problems are sorted out with different kinds of PMUs which can accumulate the low input power and then discharge to feed a battery or a device when a high enough power has been stored [75, 76, 77]. A power management system for variable input power conditions was proposed in [78]. It used AS1310 DC-DC step-up converter along with S882Z charge pump which further enhanced the performance of the DC-DC converter as shown in Figure 43 [45]. When the harvested voltage reaches higher than 0.3V oscillations start in the S882Z module and generates a clock signal. This clock signal activates the charge pump circuitry inside it and boosts the harvested voltage level. This boosted voltage is then fed to capacitor CCPOUT. When the capacitor voltage reaches a threshold level then this power is transferred to AS1310 which then supplies an appropriate power level to drive the load.

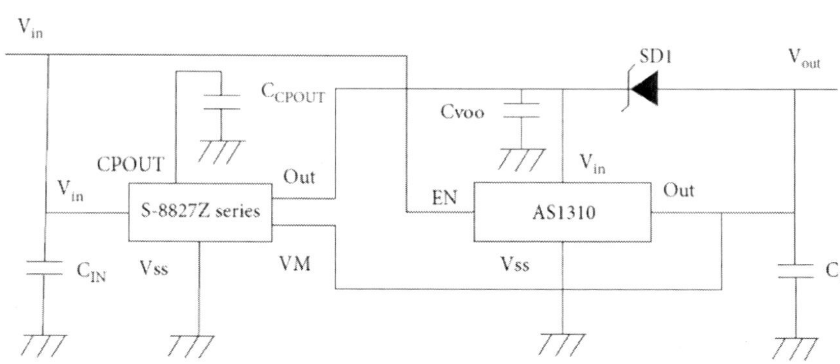

Figure 43. Power management system [45, 78].

Maximum power point tracking (MPPT) is a popular method making RFEH system robust to changes in working conditions. It works by adjusting the voltage or current level by using the V-I curve of the power source [79]. The algorithm then finds the point of maximum received power and accordingly calibrates the system for maximum power absorption. An MPPT based RF harvester was presented in [80] working over a wide range of power from −27 dBm to 6 dBm. This MPPT circuit configured the impedance matching circuit and controlled the switching of the buck-boost converter. Peak efficiency of 49% was achieved at a power level of -18 dBm. Various microcontroller-based MPPT algorithms are also present in literature [20, 81, 82, 83, 84]. A smart microcontroller-based MPPT harvester was designed in [85] as shown in Figure 44. Simulations done for input power varying from 10 µW to 1 mW showed the effectiveness of the MPPT algorithm deployed. Systems employing Discontinuous conduction mode (DCM) [86, 87] have also been used. Various RFID-specific power management systems have also been proposed in the literature for active RFID tags [88, 85].

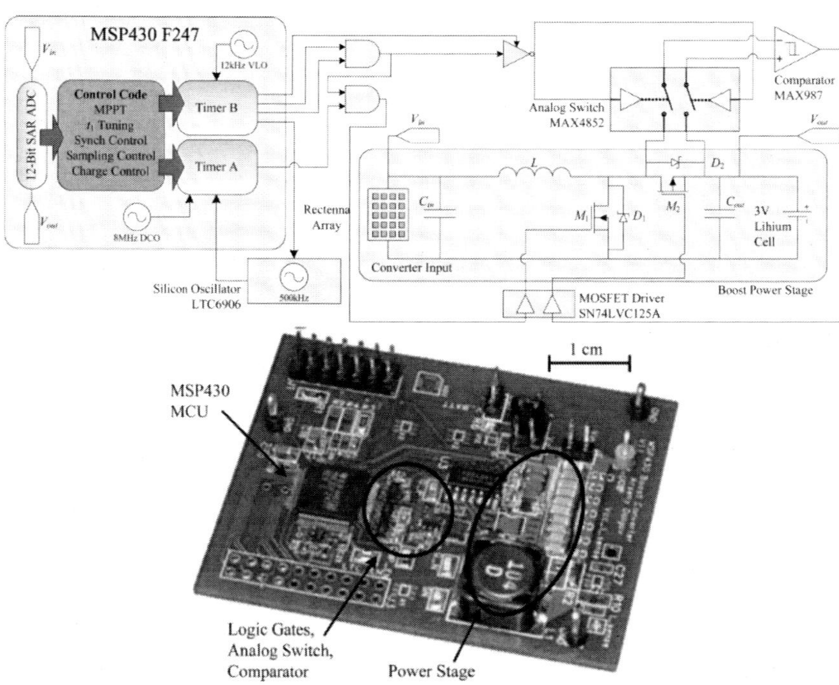

Figure 44. Microcontroller based MPPT PMU [85].

[97] used a power management system based on embedded energy-aware firmware to power peripherals connected to a 16-bit PIC24F microcontroller as shown in Figure 45. Additionally, the system used a log-periodic antenna with a maximum gain of 7.3dBi, RF-DC charge pump with a sensitivity of -14.6 dBm. With this system, a sensitivity of -14.6 dBm was achieved for the Digital TV (DTV) signals from a distance of over 6.3 km.

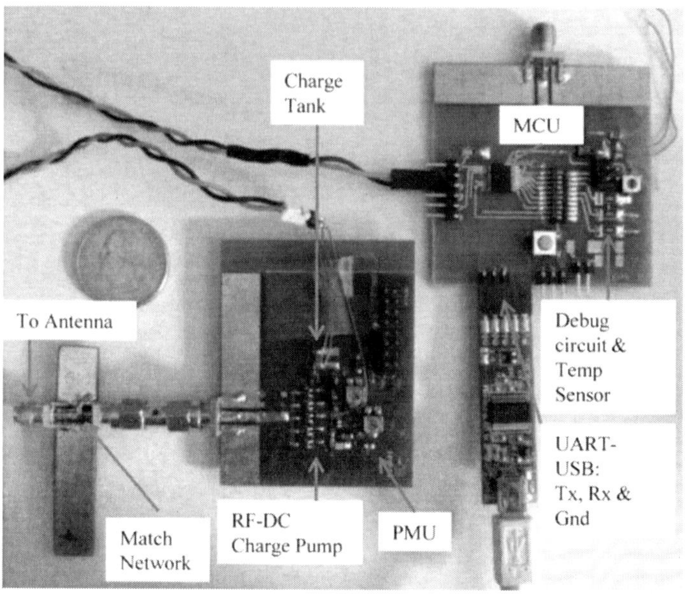

Figure 45. PMU in RFEH harvester [97].

Recent trends in computing are seeing great advancements in the field of Machine learning, an application of Artificial Intelligence. This field is now finding a place in all the different applications from all the different fields. No wonder that machine Learning is also coming into the picture of power management for RF harvesting [114, 95]. In [113] an extensive work was done on power management using Artificial Neural Networks (ANN) and near-time linear regression-based enhancer (LFNTLRE) algorithms. Sensors developed using these algorithms were used to determine optimum energy harvesting schedule and switching between the intended energy source (dedicated RF energy) and some other (ambient RF energy) source of energy depending on their availability. A hit count of 99.5% and 99.6% could be attained using ANN and LFNTLRE algorithms. The basic algorithm flow diagram is shown in Figure 46. Further incorporations of Artificial Intelligence into the design process of RF harvesting systems can bring exciting advancements in this field.

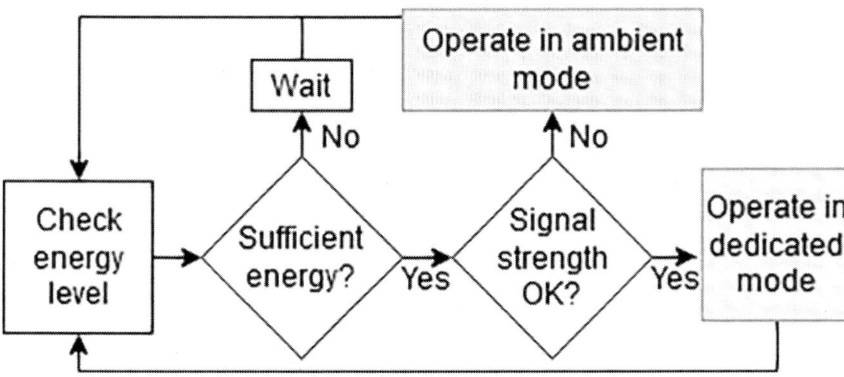

Figure 46. Algorithm for RF energy management using machine learning [113].

The following points should be deliberated over before adding a PMU to the harvesting system

- Power Management systems come with an added cost of the system, hence they must be used only when required. Marginal analysis can be carried out. However, PMU can be used for critical applications where reliability is of utmost importance.
- For the environments having a deterministic and well-defined power variation, a fixed circuit design-based PMU can perform well.
- Microcontroller-based PMU should be preferred for the environments having complex power variations or situations demanding time to time updates of the management algorithm.

8. Challenges and Limitations

So far this paper has elaborated all the different system blocks in an RF harvester. The text also shows that RF harvesting has several challenges for achieving efficient RF to DC power conversion. These include both in-system and environmental factors. The following points summarize some of the major hurdles in RF harvesting technology that have to be focused on for achieving a widespread commercial success of this budding technology.

- Power levels of the RF signals in the environment are very low because of the environmental losses and legal constraints on the transmission power. This leaves the designer with the challenge to generate a meaningful amount of energy from this feeble power source. This limits the use of RF harvesting for only very low power systems.
- Despite all the planning and design considerations, the power levels of RF waves will inevitably vary with time. As described above, efficiency is a strong function of the input power level. It disturbs the matching between antenna and rectifier. Hence changing the input power level from optimal value degrades the power conversion efficiency. The case worsens if the system is mounted on a mobile target because now the power changes with the location of the target as well. Apart from this, the direction of orientation of the antenna also changes the amount of power available to the harvester because of polarization mismatch and directivity.
- Rectifier accounts for a considerable power loss in any RF harvester. Rectifiers are inherently lossy because of the use of diodes. Diodes include losses due to parasitic components, switching speed, leakage currents, threshold, and breakdown voltage. They need optimum voltage levels to give high efficiency. The lower threshold voltage of Schottky diodes helps reduce forward drop loss but at the same time, it also leads to lower

breakdown voltage. This limits the maximum input power level which can be received from RF waves. CMOS-based rectifiers suffer from higher threshold voltage and leakage currents. Compensating the threshold voltage leads to increased leakage currents. Furthermore, the increase in frequency deteriorates CMOS efficiency due to an increase in leakage current.

- The number of stages of a rectifier/voltage multiplier directly affects the efficiency of a system. Hence an optimal number needs to be found for a balance between higher efficiencies and the output voltage level.
- Diode being a non-linear device produces harmonics of the fundamental frequency. These frequency waves travel to different parts of the circuit and cause electromagnetic interference with nearby elements, resulting in performance degradation of the whole system. Filters can be used to reject these high-frequency components to reach other components. However, doing so adds to further the design complexity of the system.
- Modern applications may need small-sized harvester which in itself is a design challenge. The miniaturization of antennas comes at the cost of gain and efficiency. Also, the smaller antennas have higher bandwidths. Because the gain-bandwidth product of an antenna is essentially a constant, miniaturization leads to lower gains. This demand for a clever trade-off between these parameters is to obtain the best possible results [107].
- In practical applications, the load connected at the output may demand different power depending on the task it might be performing. This load variation leads to a change in optimal V-I characteristics at the output and hence results in a degradation of efficiency.
- The power storage elements are needed for power backup in case of low energy in the ambient source. These elements have problems of their own, such as self-discharge and required maintenance. They also need additional charging/managing units.

Hence their inclusion also leads to the added cost of the overall system.

- Resistance compression networks (RCN) used to suppress the effect of non-linearities in the circuit are known to be very frequency sensitive and hence operate efficiently only in narrow frequency bands. Though wideband RCN networks have been reported too [137], they are mostly limited for use in broadband harvesting systems.
- There is a lack of systematic guidelines for designing an efficient system and no good theoretical limit on the harvesting efficiency has been reported. Hence setting up the different system parameters to build an efficient system becomes quite tedious. This makes the design of RF harvesting a great challenge for the designers.

CONCLUSION

RF harvesting continues to seek the great attention of researchers because of the 'free of cost and almost omnipresent nature of ambient RF radiations. In this paper, a thorough review of a typical RFEH system was presented with an elaborate discussion on different system blocks. Motivation for concentrating on this new ambient power harvesting system was understood and the developments in various technological fields were explored which have made it possible to realize such systems with practical importance. Methods adopted for making these systems compatible with the radiations present in the ambient environment were discussed. Various topologies of different system blocks were studied and compared. Insight was presented on design considerations of different system blocks. The system blocks were developed to improve the efficiency of the overall system. However, despite all the advancements, there are many challenges to be addressed. These pertaining challenges were highlighted which have to be focused on in future researches. Day by day maturing electronics to provide lower and lower losses raises hope of seeing much-improved system efficiencies in the near future. The pace of research in this field gives assurance that future researches may further help to make this field a widespread reality. The fields of embedded systems, health monitoring, WSN, RFID, and many wearable technologies will be benefitted the most and they in turn will act as a motivation to further studies for this technology. It will prove to be an excellent alternative source for a wide range of applications removing the intensive dependence on conventional power sources.

REFERENCES

[1] Brown, W. C., "Experiments involving a microwave beam to power and position a helicopter," *IEEE Trans. on Aerospace and Electronic System*, vol. AES-5, pp. 692-702, Sep. 1969.

[2] Mcspadden, J. O., and J. C. Mankins, "Space solar power programs and microwave wireless power transmission technology," *IEEE Microw.Mag.*, vol. 3, pp. 46-57, Apr. 2002.

[3] Zhang, R., and C. K. Ho, "MIMO broadcasting for simultaneous wireless information and power transfer," *IEEE Transactions on Wireless Communications*, vol. 12, no. 5, pp. 1989-2001, May 2013.

[4] Lu, Xiao, et al. "Wireless networks with RF energy harvesting: A contemporary survey." *IEEE Communications Surveys & Tutorials* 17.2 (2014): 757-789.

[5] Kumar, G., and G. Kumar. "Report on cell tower radiation submitted to Secretary, DOT, Delhi." *Electrical Engineering Dept, IIT Bombay, Powai, Mumai* (2010).

[6] Papotto, Giuseppe, et al. "A 90-nm CMOS 5-Mbps crystal-less RF-powered transceiver for wireless sensor network nodes." *IEEE Journal of Solid-State Circuits* 49.2 (2013): 335-346.

[7] Gudan, Kenneth, et al. "Ultra-low power 2.4 GHz RF energy harvesting and storage system with− 25dBm sensitivity." *2015 IEEE international conference on RFID (RFID)*. IEEE, 2015.

[8] Liu, H., "*Maximizing efficiency of wireless power transfer with resonant Inductive Coupling*," 2011. (Available on-line at http://hxhl95.github.io/media/ib ee.pdf).

[9] Kurs, A., A. Karalis, R. Moffatt, J. D. Joannopoulos, P. Fisher, and M. Soljacic, "Wireless power transfer via strongly coupled magnetic resonances," *Science*, vol. 317, no. 5834, pp. 83-86, June 2007.

[10] Pinuela Manuel, Yates David C, Lucyszyn Stepan, Mitcheson Paul D. Maximizing DC-to-load efficiency for inductive power transfer. *IEEE Trans Power Electron* 2013;28(5):2437e47.

[11] Shoki Hiroki. Issues and initiatives for practical deployment of wireless power transfer technologies in Japan. *Proc IEEE* 2013;101(6):1312e20.

[12] Cansiz, Mustafa, Dogay Altinel, and Gunes Karabulut Kurt. "Efficiency in RF energy harvesting systems: A comprehensive review." *Energy* 174 (2019): 292-309.

[13] Huang, John. "The finite ground plane effect on the microstrip antenna radiation patterns." *IEEE Transactions on Antennas and Propagation* 31.4 (1983): 649-653.

[14] Bakkali, Asma, et al. "A dual-band antenna for RF energy harvesting systems in wireless sensor networks." *Journal of Sensors* 2016 (2016).

[15] Yo, Tzong-Chee, et al. "Compact circularly polarized rectenna with unbalanced circular slots." *IEEE Transactions on Antennas and Propagation* 56.3 (2008): 882-886.

[16] Vera, Gianfranco Andia, et al. "Design of a 2.45 GHz rectenna for electromagnetic (EM) energy scavenging." *2010 IEEE Radio and Wireless Symposium (RWS)*. IEEE, 2010.

[17] Notis, Dimitris T., Phaedra C. Liakou, and Dimitris P. Chrissoulidis. "Dual polarized microstrip patch antenna, reduced in size by use of peripheral slits." *34th European Microwave Conference, 2004*. Vol. 1. IEEE, 2004.

[18] Ali, Jawad K. "A new compact size microstrip patch antenna with irregular slots for handheld GPS application." *Engineering and Technology Journal* 26.10 (2008): 1241-1246.

[19] Park, Ji-Yong, et al. *IEEE Microwave and Wireless Components Letters* 12.1 (2002): 18-19.

[20] Popović, Zoya, et al. "Scalable RF energy harvesting." *IEEE Transactions on Microwave Theory and Techniques* 62.4 (2014): 1046-1056.

[21] Sun, Hucheng, and Wen Geyi. "A new rectenna using beamwidth-enhanced antenna array for RF power harvesting applications." *IEEE Antennas and Wireless Propagation Letters* 16 (2016): 1451-1454.

[22] Hafeez, Muhammad Talha, and Syeda Fizzah Jilani. "Novel millimeter-wave flexible antenna for RF energy harvesting." *2017 IEEE International Symposium on Antennas and Propagation & USNC/URSI National Radio Science Meeting*. IEEE, 2017.

[23] Maeng, Jimin, et al. "Parylene interposer as thin flexible 3-D packaging enabler for wireless applications." *IEEE transactions on microwave theory and techniques* 59.12 (2011): 3410-3418.

[24] Adami, Salah-Eddine, et al. "A flexible 2.45-GHz power harvesting wristband with net system output from− 24.3 dBm of RF power." *IEEE Transactions on Microwave Theory and Techniques* 66.1 (2017): 380-395.

[25] Agrawal, Sachin, et al. "An efficient RF energy harvester with tuned matching circuit." *VLSI Design and Test*. Springer, Berlin, Heidelberg, 2013. 138-145.

[26] Al-Azawy, Musaab Mohammed, and Filiz Sari. "Analysis of Dickson Voltage Multiplier for RF Energy Harvesting." *2019 1st Global Power, Energy and Communication Conference (GPECOM)*. IEEE, 2019.

[27] Penella-López, María Teresa, and Manuel Gasulla-Forner. "Radiofrequency energy harvesting." *Powering Autonomous Sensors*. Springer, Dordrecht, 2011. 125-147.

[28] Merz, Christian, and Gerald Kupris. "High Q impedance matching for RF energy harvesting applications." *2016 3rd International Symposium on Wireless Systems within the Conferences on*

Intelligent Data Acquisition and Advanced Computing Systems (IDAACS-SWS). IEEE, 2016.

[29] Hameed, Zohaib, and Kambiz Moez. "Design of impedance matching circuits for RF energy harvesting systems." *Microelectronics Journal* 62 (2017): 49-56.

[30] Agrawal, Sachin, et al. "Realization of efficient RF energy harvesting circuits employing different matching technique." *Fifteenth International Symposium on Quality Electronic Design*. IEEE, 2014.

[31] Scucchia, Lucio, et al. "Reconfigurable matching network for RF energy harvesting circuits." *2014 International Workshop on Integrated Nonlinear Microwave and Millimetre-wave Circuits (INMMiC)*. IEEE, 2014.

[32] bin Abdul Aziz, Muhammad Ezanuddin, et al. "*Rectifier for RF energy harvesting using stub matching*." (2019).

[33] Kimionis, John, et al. "Octave and decade printed UWB rectifiers based on nonuniform transmission lines for energy harvesting." *IEEE Transactions on Microwave Theory and Techniques* 65.11 (2017): 4326-4334.

[34] Ungan, Tolgay, et al. "RF energy harvesting design using high Q resonators." *2009 IEEE MTT-S International Microwave Workshop on Wireless Sensing, Local Positioning, and RFID*. IEEE, 2009.

[35] Barnett, Raymond E., Jin Liu, and Steve Lazar. "A RF to DC voltage conversion model for multi-stage rectifiers in UHF RFID transponders." *IEEE Journal of solid-state circuits* 44.2 (2009): 354-370.

[36] Stoopman, Mark, et al. "Co-design of a CMOS rectifier and small loop antenna for highly sensitive RF energy harvesters." *IEEE Journal of Solid-State Circuits* 49.3 (2014): 622-634.

[37] Hameed, Zohaib, and Kambiz Moez. "Design of impedance matching circuits for RF energy harvesting systems." *Microelectronics Journal* 62 (2017): 49-56.

[38] Zhao, Ping, Thomas Hollstein, and Manfred Glesner. "Analysis on power harvesting circuits with tunable matching network for

improved efficiency." 2009 PhD *Research in Microelectronics and Electronics*. IEEE, 2009.

[39] Hameed, Zohaib, and Kambiz Moez. "Design of impedance matching circuits for RF energy harvesting systems." *Microelectronics Journal* 62 (2017): 49-56.

[40] Soltani, N., & Yuan, F. (2010). A high-gain power-matching technique for efficient radio-frequency power harvest of passive wireless microsystems. *IEEE Transactions on Circuits and Systems I: Regular Papers,* 57(10), 2685–2695.

[41] Soltani, N., & Tuan, F. (2010). A step-up transformer impedance transformation technique for efficient power harvesting of passive transponders. *Microelectronics Journal,* 41(2010), 75–84.

[42] Scucchia, Lucio, et al. "Reconfigurable matching network for RF energy harvesting circuits." *2014 International Workshop on Integrated Nonlinear Microwave and Millimetre-wave Circuits (INMMiC)*. IEEE, 2014.

[43] Zhu, Ning, et al. "Design of a high-efficiency rectenna for 1.575 GHz wireless low power transmission." *2011 IEEE Radio and Wireless Symposium*. IEEE, 2011.

[44] Filiz, S. A. R. I., and U. Z. U. N. Yunus. "A Comparative Study: Voltage Multipliers for RF Energy Harvesting System." *Communications Faculty of Sciences University of Ankara Series A2-A3 Physical Sciences and Engineering* 61.1 (2019): 12-23.

[45] Saeed, Warda, et al. "RF energy harvesting for ubiquitous, zero power wireless sensors." *International Journal of Antennas and Propagation* 2018 (2018).

[46] Loo, Xi Sung, et al. "High Voltage Energy Harvesters." *Advanced Electronic Circuits-Principles, Architectures and Applications on Emerging Technologies*. IntechOpen, 2017.

[47] Marshall, Blake R., Marcin M. Morys, and Gregory D. Durgin. "Parametric analysis and design guidelines of RF-to-DC Dickson charge pumps for RFID energy harvesting." *2015 IEEE International Conference on RFID (RFID)*. IEEE, 2015.

[48] Dickson, John F. "On-chip high-voltage generation in MNOS integrated circuits using an improved voltage multiplier technique." *IEEE Journal of solid-state circuits* 11.3 (1976): 374-378.

[49] Barnett, Raymond E., Jin Liu, and Steve Lazar. "A RF to DC voltage conversion model for multi-stage rectifiers in UHF RFID transponders." *IEEE Journal of solid-state circuits* 44.2 (2009): 354-370.

[50] Dallago, Enrico, et al. "Self-supplied integrable active high-efficiency AC-DC converter for piezoelectric energy scavenging systems." *2007 IEEE International Symposium on Circuits and Systems*. IEEE, 2007.

[51] Tran, Le-Giang, Hyouk-Kyu Cha, and Woo-Tae Park. "RF power harvesting: a review on designing methodologies and applications." *Micro and Nano Systems Letters* 5.1 (2017): 14.

[52] Almorabeti, Soumaya, Mounir Rifi, and Hanae Terchoune. "*Rectifier Circuit Designs for RF Energy Harvesting applications*." 2019.

[53] Taybi, Abdellah, et al. "Design of a new 5.8 GHz RF-DC rectifier structure for wireless power transmission." *2017 International Conference on Wireless Technologies, Embedded and Intelligent Systems (WITS)*. IEEE, 2017.

[54] Li, Cheng-Jyun, and Tai-Cheng Lee. "2.4-GHz high-efficiency adaptive power." *IEEE Transactions on Very Large Scale Integration (VLSI) Systems* 22.2 (2013): 434-438.

[55] Mendoza Domínguez, Joan Manuel. *On the design of a RF schottky diode rectifier for energy harvesting applications*. BS thesis. Quito, 2018.

[56] Partal, Hakan P., Mehmet A. Belen, and Sibel Zorlu Partal. "Design and realization of an ultra-low power sensing RF energy harvesting module with its RF and DC sub-components." *International Journal of RF and Microwave Computer-Aided Engineering* 29.1 (2019): e21622.

[57] Melais, Sergio E. "*Design and optimization of broadband planar baluns and dipole antennas*." (2005).

[58] Le, Triet, Karti Mayaram, and Terri Fiez. "Efficient far-field radio frequency energy harvesting for passively powered sensor networks." *IEEE Journal of solid-state circuits* 43.5 (2008): 1287-1302.

[59] Serdijn, W. A., A. L. R. Mansano, and M. Stoopman. "Introduction to RF energy harvesting." *Wearable Sensors*. Academic Press, 2014. 299-322.

[60] Le, Triet, Karti Mayaram, and Terri Fiez. "Efficient far-field radio frequency energy harvesting for passively powered sensor networks." *IEEE Journal of solid-state circuits* 43.5 (2008): 1287-1302.

[61] Lin, Hongchin, Kai-Hsun Chang, and Shyh-Chyi Wong. "Novel high positive and negative pumping circuits for low supply voltage." *1999 IEEE International Symposium on Circuits and Systems (ISCAS)*. Vol. 1. IEEE, 1999.

[62] Papotto, Giuseppe, Francesco Carrara, and Giuseppe Palmisano. "A 90-nm CMOS threshold-compensated RF energy harvester." *IEEE Journal of Solid-State Circuits* 46.9 (2011): 1985-1997.

[63] Nakamoto, Hiroyuki, et al. "A Passive UHF RF Identification CMOS Tag IC Using Ferroelectric RAM in 0.35-$\mu {\hbox {m}} $ Technology." *IEEE Journal of Solid-State Circuits* 42.1 (2006): 101-110.

[64] Mansano, Andre, Sumit Bagga, and Wouter Serdijn. "A high efficiency orthogonally switching passive charge pump rectifier for energy harvesters." *IEEE Transactions on Circuits and Systems I: Regular Papers* 60.7 (2013): 1959-1966.

[65] Hwang, Yuh-Shyan, et al. "A 13.56-MHz low-voltage and low-control-loss RF-DC rectifier utilizing a reducing reverse loss technique." *IEEE Transactions on Power Electronics* 29.12 (2014): 6544-6554.

[66] Lehmann, Torsten, and Yashodhan Moghe. "On-chip active power rectifiers for biomedical applications." *2005 IEEE International Symposium on Circuits and Systems*. IEEE, 2005.

[67] Dallago, Enrico, et al. "Self-supplied integrable active high-efficiency AC-DC converter for piezoelectric energy scavenging systems." *2007 IEEE International Symposium on Circuits and Systems*. IEEE, 2007.

[68] Cheng, Shuo, et al. "An active voltage doubling AC/DC converter for low-voltage energy harvesting applications." *IEEE Transactions on Power Electronics* 26.8 (2010): 2258-2265.

[69] Sun, Yang, et al. "An integrated high-performance active rectifier for piezoelectric vibration energy harvesting systems." *IEEE Transactions on Power Electronics* 27.2 (2011): 623-627.

[70] Prauzek, Michal, et al. "Energy harvesting sources, storage devices and system topologies for environmental wireless sensor networks: A review." *Sensors* 18.8 (2018): 2446.

[71] Zhang, Fan, et al. "A batteryless 19μW MICS/ISM-band energy harvesting body area sensor node SoC." *2012 IEEE International Solid-State Circuits Conference*. IEEE, 2012.

[72] Gudan, Kenneth, et al. "A 2.4 GHz ambient RF energy harvesting system with− 20dBm minimum input power and NiMH battery storage." *2014 IEEE RFID Technology and Applications Conference (RFID-TA)*. IEEE, 2014.

[73] Nintanavongsa, Prusayon, et al. "Design optimization and implementation for RF energy harvesting circuits." *IEEE Journal on emerging and selected topics in circuits and systems* 2.1 (2012): 24-33.

[74] Dolgov, Arseny, Regan Zane, and Zoya Popovic. "Power management system for online low power RF energy harvesting optimization." *IEEE Transactions on Circuits and Systems I: Regular Papers* 57.7 (2010): 1802-1811.

[75] Abouzied, Mohamed A., Krishnan Ravichandran, and Edgar Sánchez-Sinencio. "A fully integrated reconfigurable self-startup RF energy-harvesting system with storage capability." *IEEE Journal of Solid-State Circuits* 52.3 (2016): 704-719.

[76] Safarian, Zahra, and Hossein Hashemi. "Wirelessly powered passive systems with dynamic energy storage mechanism." *IEEE*

transactions on microwave theory and techniques 62.4 (2014): 1012-1021.

[77] Reinisch, Hannes, et al. "An electro-magnetic energy harvesting system with 190 nW idle mode power consumption for a BAW based wireless sensor node." *IEEE Journal of solid-state circuits* 46.7 (2011): 1728-1741.

[78] Olgun, Ugur, CC. Chen, and John L. Volakis. "Design of an efficient ambient WiFi energy harvesting system." *IET Microwaves, Antennas & Propagation* 6.11 (2012): 1200-1206.

[79] Kansal, Aman, et al. "Harvesting aware power management for sensor networks." *Proceedings of the 43rd annual Design Automation Conference*. 2006.

[80] Martins, Gustavo C., and Wouter A. Serdijn. "An RF energy harvester with MPPT operating across a wide range of available input power." *2018 IEEE international symposium on circuits and systems (ISCAS)*. IEEE, 2018.

[81] Paing, Thurein, et al. "Resistor emulation approach to low-power RF energy harvesting." *IEEE Transactions on Power Electronics* 23.3 (2008): 1494-1501.

[82] Costinett, Daniel, et al. "RF-powered variable duty cycle wireless sensor." *The 40th European Microwave Conference*. IEEE, 2010.

[83] Falkenstein, Erez, et al. "Far-field RF-powered variable duty cycle wireless sensor platform." *IEEE Transactions on Circuits and Systems II: Express Briefs* 58.12 (2011): 822-826.

[84] Paing, T., et al. "Wirelessly-powered wireless sensor platform." *2007 European Microwave Conference*. IEEE, 2007.

[85] Dolgov, Arseny, Regan Zane, and Zoya Popovic. "Power management system for online low power RF energy harvesting optimization." *IEEE Transactions on Circuits and Systems I: Regular Papers* 57.7 (2010): 1802-1811.

[86] Paing, T., J. Shin, R. Zane, and Z. Popovic, "Resistor emulation approachto low-power RF energy harvesting," *IEEE Trans. Power Electron.*, vol. 23, no. 3, pp. 1494–1501, May 2008.

[87] Paing, T., J. Morroni, A. Dolgov, J. Shin, J. Brannan, R. Zane, and Z. Popovic, “Wirelessly-powered wireless sensor platform,” in *2007 European Conf. Wireless Technologies*, Oct. 8–10, 2007, pp. 241–244.

[88] Janek, Alex, et al. "Power management strategies for battery-driven higher Class UHF RFID tags supported by energy harvesting devices." *2007 IEEE Workshop on Automatic Identification Advanced Technologies*. IEEE, 2007.

[89] https://batteryuniversity.com/learn/article/whats_the_role_of_the_supercapacitor#:~:text=Lithium%2Dion%20batteries%20share%20a,times%20less%20than%20Li%2Dion.&text=A%20battery%20with%20a%20flat,before%20reaching%20the%20voltage%20threshold.

[90] Stevenson, Anthony J., et al. "Supercapatteries with hybrids of redox active polymers and nanostructured carbons." *Nanocarbons for Advanced Energy Storage* 1 (2015): 179-210.

[91] Klumpner, Christian, Greg Asher, and George Z. Chen. "Selecting the power electronic interface for a supercapattery based energy storage system." *2009 IEEE Bucharest PowerTech*. IEEE, 2009.

[92] Libich, J., Máca, J., Vondrák, J., Čech, O., Sedlaříková, M. Supercapacitors: Properties and applications. *J. Energy Storage* 2018, 17, 224–227. [CrossRef].

[93] Renner, C., Jessen, J., Turau, V. Lifetime prediction for supercapacitor-powered wireless sensor nodes. In *Proceedings of the GI/ITG Fachgespräch “Sensornetze” (FGSN’09)*, Humburg, Germany, 13 August 2009; pp. 1–6.

[94] Amanor-Boadu, Judy, et al. "A switched mode Li-ion battery charger with multiple energy harvesting systems simultaneously used as input sources." *2014 IEEE 57th International Midwest Symposium on Circuits and Systems (MWSCAS)*. IEEE, 2014.

[95] Chu, M., H. Li, X. Liao and S. Cui, "Reinforcement learning-based multiaccess control and battery prediction with energy harvesting in IoT systems," *IEEE Internet Things J.*, vol. 6, no. 2, pp. 2009-2020, Apr. 2019.

[96] Sadiku, Matthew NO. *Elements of electromagnetics*. Oxford university press, 2014.

[97] Vyas, Rushi J., et al. "E-WEHP: A batteryless embedded sensor-platform wirelessly powered from ambient digital-TV signals." *IEEE Transactions on microwave theory and techniques* 61.6 (2013): 2491-2505.

[98] Gonçalves, Hugo, Miguel Martins, and Jorge Fernandes. "Fully integrated energy harvesting circuit with− 25-dBm sensitivity using transformer matching." *IEEE Transactions on Circuits and Systems II: Express Briefs* 62.5 (2015): 446-450.

[99] Sood, Er Vishakha, and Er Manwinder Singh. "Spectrum Utilization by Using Cognitive Radio Technology." *International Journal of Advanced Engineering Sciences and Technologies, Volume* 7: 258-263.

[100] Harish, A. R., and M. Sachidananda. *Antennas and wave propagation*. Oxford University Press, USA, 2007.

[101] Raju, G. S. N. *Antennas and wave propagation*. Pearson Education India, 2006.

[102] Pozar, David M. *Microwave engineering*. John Wiley & Sons, 2009.

[103] Ruzbehani, Mohsen. "A Comparative Study of Symmetrical Cockcroft-Walton Voltage Multipliers." *Journal of Electrical and Computer Engineering* 2017 (2017).

[104] Schantz, Hans Gregory. "Introduction to ultra-wideband antennas." *IEEE Conference on Ultra Wideband Systems and Technologies, 2003*. IEEE, 2003.

[105] Deschamps, G., and P. Mast. "Poincaré sphere representation of partially polarized fields." *IEEE Transactions on Antennas and Propagation* 21.4 (1973): 474-478.

[106] Mikeka, C., and H. Arai. "Notched Circular Microstrip Patch Antenna integrated with a single diode rectifier for energy-harvesting prosthetic leg." *2011 IEEE Topical Conference on Biomedical Wireless Technologies, Networks, and Sensing Systems*. IEEE, 2011.

[107] Divakaran, Sleebi K., and Deepti Das Krishna. "RF energy harvesting systems: An overview and design issues." *International Journal of RF and Microwave Computer-Aided Engineering* 29.1 (2019): e21633.

[108] Vullers, R. J. M., et al. "Micropower energy harvesting." *Solid-State Electronics* 53.7 (2009): 684-693.

[109] Akhtar, Fayaz, and Mubashir Husain Rehmani. "Energy replenishment using renewable and traditional energy resources for sustainable wireless sensor networks: A review." *Renewable and Sustainable Energy Reviews* 45 (2015): 769-784.

[110] Yoo, T. W., and Kai Chang. "Theoretical and experimental development of 10 and 35 GHz rectennas." *IEEE Transactions on Microwave Theory and Techniques* 40.6 (1992): 1259-1266.

[111] Hong, Haocheng, et al. "Demonstration of a highly efficient RF energy harvester for Wi-Fi signals." *2012 International Conference on Microwave and Millimeter Wave Technology (ICMMT)*. Vol. 5. IEEE, 2012.

[112] Bhawan, Mahanagar Doorsanchar, and Jawahar Lal Nehru Marg. "Effects of Electromagnetic Field Radiation from Mobile Towers and Handsets." *Telecom Regulatory Authority of India* (2014).

[113] Kwan, Jonathan C., Jesse M. Chaulk, and Abraham O. Fapojuwo. "A Coordinated Ambient/Dedicated Radio Frequency Energy Harvesting Scheme using Machine Learning." *IEEE Sensors Journal* (2020).

[114] Sharma, M. K., A. Zappone, M. Assaad, M. Debbah and S. Vassilaras, "Distributed power control for large energy harvesting networks: A multi-agent deep reinforcement learning approach," *IEEE Trans. Cognit. Commun. Netw.*, vol. 5, no. 4, pp. 1140-1154, Dec. 2019.

[115] Zhang, Xu, et al. "Opportunities and Challenges of Ambient Radio-Frequency Energy Harvesting." *Joule* (2020).

[116] Kwan, J. C., and Fapojuwo, A. O. (2016).Measurement and Analysis of Available Ambient Radio Frequency Energy for

Wireless Energy Harvesting. In *2016 IEEE 84th Vehicular Technology Conference* (VTC-Fall), pp. 1–6.

[117] Li, X., Dennis Heo, U., Ma, K., Narayanan, V., Liu, H., and Datta, S. (2014). RF-powered systems using steep-slope devices. In 2014 *IEEE 12th International New Circuits and Systems Conference* (NEWCAS), pp. 73–76.

[118] Oh, S., Blaauw, D., and Sylvester, D. (2019). The Internet of Tiny Things: Recent Advances of Millimeter-Scale Computing. *IEEE Des. Test* 36, 65–72.

[119] Elwi, Taha A., Daniya Amer Jassim, and Hayder Hassan Mohammed. "Novel miniaturized folded UWB microstrip antenna-based metamaterial for RF energy harvesting." *International Journal of Communication Systems* 33.6 (2020): e4305.

[120] Mansour, Mohamed M., Shuya Yamamoto, and Haruichi Kanaya. "Reconfigurable Multistage RF Rectifier Topology for 900 MHz ISM Energy-Harvesting Applications." *IEEE Microwave and Wireless Components Letters* (2020).

[121] Estrada, José Antonio, et al. "RF-Harvesting Tightly Coupled Rectenna Array Tee-Shirt With Greater Than Octave Bandwidth." *IEEE Transactions on Microwave Theory and Techniques* (2020).

[122] Amer, Abdulrahman Ahmed Ghaleb, et al. "A Comprehensive Review of Metasurface Structures Suitable for RF Energy Harvesting." *IEEE Access* 8 (2020): 76433-76452.

[123] Piñuela, Manuel, Paul D. Mitcheson, and Stepan Lucyszyn. "Ambient RF energy harvesting in urban and semi-urban environments." *IEEE Transactions on microwave theory and techniques* 61.7 (2013): 2715-2726.

[124] Shen, Shanpu, et al. "An ambient RF energy harvesting system where the number of antenna ports is dependent on frequency." *IEEE Transactions on Microwave Theory and Techniques* 67.9 (2019): 3821-3832.

[125] Wagih, Mahmoud, Alex S. Weddell, and Steve Beeby. "Rectennas for Radio-Frequency Energy Harvesting and Wireless Power Transfer: A Review of Antenna Design [Antenna Applications

Corner]." *IEEE Antennas and Propagation Magazine* 62.5 (2020): 95-107.

[126] Dai, Haojuan, et al. "A review and design of the on-chip rectifiers for RF energy harvesting." *2015 IEEE International Wireless Symposium (IWS 2015)*. IEEE, 2015.

[127] Soyata, Tolga, Lucian Copeland, and Wendi Heinzelman. "RF energy harvesting for embedded systems: A survey of tradeoffs and methodology." *IEEE Circuits and Systems Magazine* 16.1 (2016): 22-57.

[128] Chong, Gabriel, et al. "Ambient RF energy harvesting system: a review on integrated circuit design." *Analog Integrated Circuits and Signal Processing* 97.3 (2018): 515-531.

[129] Baroudi, Uthman. "Management of RF Energy Harvesting: A Survey." *2019 16th International Multi-Conference on Systems, Signals & Devices (SSD)*. IEEE, 2019.

[130] Visser, Hubregt J., and Ruud JM Vullers. "RF energy harvesting and transport for wireless sensor network applications: Principles and requirements." *Proceedings of the IEEE* 101.6 (2013): 1410-1423.

[131] Lopez, Alfred R. "Review of narrowband impedance-matching limitations." *IEEE Antennas and Propagation Magazine* 46.4 (2004): 88-90.

[132] Sun, Yichuang, and J. K. Fidler. "Design of/spl Pi/impedance matching networks." *Proceedings of IEEE International Symposium on Circuits and Systems-ISCAS'94*. Vol. 5. IEEE, 1994.

[133] Nimo, Antwi, et al. "Analysis of passive RF-DC power rectification and harvesting wireless RF energy for micro-watt sensors." *AIMS Energy* 3.2 (2015): 184-200.

[134] Fan, Shiquan, et al. "A 2.45-GHz rectifier-booster regulator with impedance matching converters for wireless energy harvesting." *IEEE Transactions on Microwave Theory and Techniques* 67.9 (2019): 3833-3843.

[135] Song, Chaoyun, et al. "Matching network elimination in broadband rectennas for high-efficiency wireless power transfer and energy

harvesting." *IEEE Transactions on Industrial Electronics* 64.5 (2016): 3950-3961.

[136] Chen, Yen-Sheng, and Cheng-Wei Chiu. "Maximum achievable power conversion efficiency obtained through an optimized rectenna structure for RF energy harvesting." *IEEE Transactions on Antennas and Propagation* 65.5 (2017): 2305-2317.

[137] Song, Chaoyun, et al. "Recent advances in broadband rectennas for wireless power transfer and ambient RF energy harvesting." *2017 11th European Conference on Antennas and Propagation (EUCAP)*. IEEE, 2017.

[138] Du, Zhi-Xia, and Xiu Yin Zhang. "High-efficiency single-and dual-band rectifiers using a complex impedance compression network for wireless power transfer." *IEEE Transactions on industrial electronics* 65.6 (2017): 5012-5022.

[139] Han, Yehui, et al. "Resistance compression networks for radio-frequency power conversion." *IEEE Transactions on Power Electronics* 22.1 (2007): 41-53.

[140] Lin, Quan Wei, and Xiu Yin Zhang. "Differential rectifier using resistance compression network for improving efficiency over extended input power range." *IEEE Transactions on Microwave Theory and Techniques* 64.9 (2016): 2943-2954.

[141] Du, Zhi-Xia, and Xiu Yin Zhang. "High-efficiency single-and dual-band rectifiers using a complex impedance compression network for wireless power transfer." *IEEE Transactions on industrial electronics* 65.6 (2017): 5012-5022.

[142] Shrestha, Sika, Sun-Kuk Noh, and Dong-You Choi. "Comparative study of antenna designs for RF energy harvesting." *International Journal of Antennas and Propagation* 2013 (2013).

[143] Cavalheiro, David, Francesc Moll, and Stanimir Valtchev. "Tunnel FET device characteristics for RF energy harvesting passive rectifiers." *2015 IEEE 13th International New Circuits and Systems Conference (NEWCAS)*. IEEE, 2015.

[144] Li, Xueqing, et al. "RF-powered systems using steep-slope devices." *2014 IEEE 12th International New Circuits and Systems Conference (NEWCAS)*. IEEE, 2014.

[145] Khantwal, Bhuvnesh, and Reeta Verma. "A Meta-evaluation of Components of RF Energy Harvesting System." *2021 8th International Conference on Computing for Sustainable Global Development (INDIACom)*. IEEE, 2021.

[146] Kharchenko, Valeriy, et al. "Determination of the Efficiency of Photovoltaic Converters Adequate to Solar Radiation by Using Their Spectral Characteristics." *International Conference on Intelligent Computing & Optimization*. Springer, Cham, 2020.

PART B: TRANSPORTATION INDUSTRY

In: Advances of Machine Learning … ISBN: 978-1-68507-211-7
Editors: Pandian Vasant et al.

Chapter 8

THE MANAGEMENT OF COMMUNITY PARTICIPATION IN RURAL INFRASTRUCTURE DEVELOPMENT IN THE MEKONG RIVER DELTA, VIETNAM

***Nguyen Xuan Quyet*[1,*], *Pham Thi My Dung*[2], *Duc Anh Nguyen*[3], *Dinh Tuan Hai*[4,†] *and Nguyen Viet Luan*[5]**

[1]Hochiminh City's University of Food and Industry, Vietnam
[2]Science Institute of Rural Development, Vietnam
[3]Faculty of Department of Building and Industrial Construction, Hanoi University of Civil Engineering, Vietnam
[4]Faculty of Civil Engineering, Hanoi Architectural University, Vietnam
[5]Faculty of Economics, South Can Tho University, Vietnam

[*] Corresponding Author's Email: quyetan25@yahoo.com.
[†] Corresponding Author's Email: haidt@hau.edu.vn.

ABSTRACT

The total length of rural roads in Vietnam, including district roads, communal roads, and village roads, has increased to 272,861 km, accounting for 82% of the length of the road network. A short-term and reasonable investment to the road network is estimated at a cost of $6.6 billion. This is a very large number, considering the state budget is limited. This enquires increasing community contribution to fulfil the needs. Theoretically, construction projects with capital contributions from the public must ensure the participation of the public in all phases of the projects. However, government policies and management mechanisms have shown a lack of transparency, leading to law abuse and misdemeanor in local project management. This study reviews the real situation of community participation in rural infrastructure development in Vietnam until 2020 by a careful investigation of laws, policies, and an analysis of questionnaires collected from the Mekong river delta. This study then proposes a few approaches to improve the participation of community and its effectiveness in rural infrastructure development – the most important ones include the construction of mechanisms of the participation and training the communities.

INTRODUCTION

According to the Vietnamese government's statistics (MoT 2018), the infrastructure development program has attracted a large amount of investment. One of the most important sectors – the rural infrastructure development has claimed $14 billion until 2019. To increase investments to the rural infrastructure, policies relating to the participation of public (i.e., organization, management, and supervision of the investment activities) need to be passed so that the government can take advantage of the public resources (Khanh 2019).

Some negative observations of the rural transport development have been recorded. These negative observations include losses and wastes in financial resources and low quality of the buildings. Whereas the participation of community has been limited: the management of these projects have been assigned mainly by the local governments. Potential causes of the limited community participation include the background of the community widely ranges and the education level of the community is low. Therefore, people in the community who want to participate in local infrastructure development should be trained regularly (An and Tung 2016).

This research is based on a systematic approach, institutional approach, policy approach, and participatory approach to propose solutions and models to enhance the effectiveness of the management of community participation in rural transport development. Its scope is rural transport infrastructure in the Mekong Delta Provinces in the Southern of Vietnam.

Data used in this study include officially published secondary data, research works, and related assessment reports of ministries, departments, sectors, and related agencies: a large part was from the Ministry of Transport and the Ministry of Agriculture and Rural Development, local departments and agencies and representatives of the community. The research method is mainly qualitative with community interviews and questionnaires obtained from 120 community representatives in the research area and from the experts who are managers and in-depth researchers. Results of the questionaires are analyzed with quantitative methods.

COMMUNITY PARTICIPATION IN RURAL INFRASTRUCTURE DEVELOPMENT

Community and Community Participation

Community

Community was defined as early as in Toennies' (1887): "groupings based on feelings of togetherness and on mutual bonds, which are felt as a goal to be kept up, their members being means for this goal" The characteristics of community are, considered by Park (1936): "The essential characteristics of a community, so conceived, are those of: (1) a population territorially organized, (2) more or less completely rooted in the soil it occupies, (3) its individual units living in a relationship of mutual interdependence". Simply put, a community includes members who have common benefits. In this research, the common benefits are the right to participate in the development of infrastructure and the ability to consume the services of the infrastructure. This induces to a review of community participation.

Community Participation

Community participation (or public participation) is defined as:

> "Public participation is based on the belief that people whose lives are affected by transportation planning and investment decisions have a right to be involved in the decision-making process and influence choices that are made. Directly engaging citizens in this process promotes successful problem solving, yields diverse voices and new ideas, and gives the public a sense of ownership of the developed solutions" (MoC, 2021).

The participation of community in building infrastructure can be considered a citizen-centered approach of infrastructure development, which implies higher effectiveness of the public governance (Bellone et al. 2018).

Less formally, community participation, is referred as "the social process of taking part (voluntarily) in formal or informal activities, programs and/or discussions to bring about a planned change or improvement in community life, services and/or resources" (Bac Lieu 2017). Therefore, volunteerism is a crucial part of the community participation concept. Through community participation, individuals in communities can maximize their potential and progress from individual action to collective social and political change (ADB 2020). Community participation adds many benefits to project development in general (Clanrence 1961).

Rural Transport Infrastructure Development

Rural Transport Infrastructure

In Vietnam, the rural transport infrastructure is a part of the transport infrastructure. Specifically, the road rural transport infrastructure serves rural internal travel and is linked with other infrastructure systems to develop products and serve economic, cultural, and social exchanges of localities. The subsystem of roads in the rural transport infrastructure includes a network of roads and bridges at district, commune, village, and lane levels, serving rural transport, directly serving the transportation of goods and the exchange and travel of residents. This system ensures that medium, light, and rudimentary motor vehicles can move along the whole network.

Rural Transport Infrastructure Development

The government of Vietnam focuses on the concept of sustainability in developing infrastructure, adopting a definition by Gustav and Thomas (1997), which is considered by themselves "creatively ambiguous": "humanity has the ability to make development sustainable – to ensure that it meets the needs of the present without compromising the ability of future generations to meet their own needs".

Therefore, the authors set the point-of-departure that rural transport infrastructure development should have the following characteristics:

1. It is a type of development with synchronous or discrete development of rural transport infrastructure in an area or a locality, in a certain period, related to roads, waterways, bridges, ports, wharfs, yards, etc., and the attached infrastructure components such as traffic lights, signs, lighting, and supply system drainages.
2. Rural infrastructure developments can be distinguished by their scopes and scales, which are the development of the number of rural transport infrastructure in a region, a locality in a certain period related to the number of types of infrastructure, the amount of construction investment capital, or the number of works or work items. The scale of development is often associated with projects or works with categories according to time and progress of implementation (e.g., estimation, on-going, completed and put into use), by size and quantity of capital (large, medium, and small), and by the level of management and function (state level, district level, village level, and lane level).

3. Rural infrastructure developments can also be distinguished by the quality of development – the level of the improvement of the efficiency of the planning, construction, and use, management, and maintenance of rural transport infrastructure. The level of improvement is connected with the socio-economic efficiency when the development satisfies the travel needs of the community and creates the development of trade.
4. Sustainable rural transport infrastructure development is the participatory management and development of rural transport infrastructure to ensure harmony between the social needs and increasing economic trade of the local community. This view is in line with the country's socio-economic development strategy in an immediate and long-term orientation.

Community Participation in Rural Transport Infrastructure Development

Forms of Community Participation in Rural Transport Infrastructure Development

In the early years of reform of respectful management, the state government, when considering the process of community participation in rural transport infrastructure development, practiced the motto "People know, people discuss, people do, people check" (Hung 2011). Quyen (2015) drew practical experience and expanded the slogan "people contribute, people manage, and people benefit", demonstrating the fairness and sustainability of community participation in the rural transport infrastructure development. Previously, a study by the Center for Rural Development in Central Vietnam from 1995 to 2005 on infrastructure development with community participation showed that all approaches to rural transport infrastructure development would start from the people, "All decisions of the people should be associated with the efficiency and benefits that each individual in the community enjoys, the project is conducted according to the motto 'People know, people discuss, people do, people check and people enjoy the benefits" (Hien 2006).

From these points of view, the authors believe that to increase the number of participants and maximize the capacity of community participation, the form of community participation development in rural transport infrastructure development should enclose the aspects that are depicted in Figure 1.

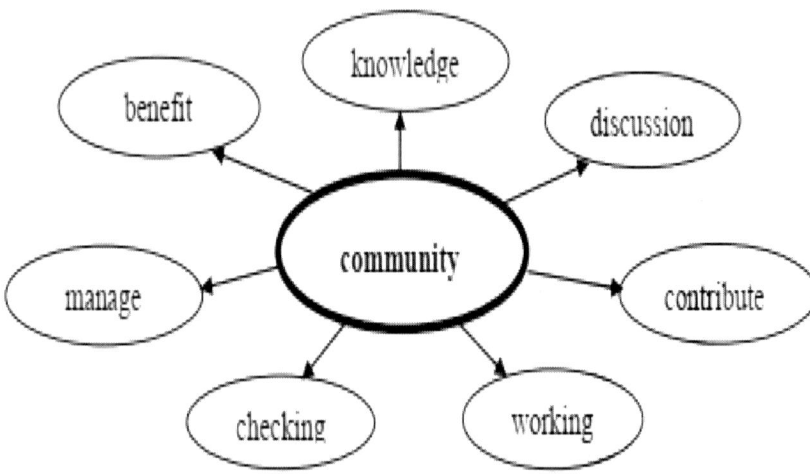

Figure 1. Aspects (forms) of community participation in rural transport infrastructure development – Adapted from Hien (2006), Hung (2011), and Quyen (2015).

In summary, community participation in rural transport infrastructure development, is a critical need with diverse forms of participation. Depending on the specific conditions of each region/locality and the ability of the community to participate, the form of community participation may vary or be limited. Specifically, with well-developed local conditions, the ability to contribute material resources is better with the participation of many economic sectors and subjects; the contributions are concerned with better material, intellectual, and financial resources. On the contrary, localities with underdeveloped socio-economic conditions often limit contributions. In addition, community management capacity, management model, and participation demands also contribute to the diversity of community participation in infrastructure development.

The Role of Community Participation in Rural Transport Infrastructure Development

Bellone and Geropanta (2018) subjected that successful infrastructure development depends on three pillars: territorial planning, technology, and social practices. The first and the third pillars implied deeper participation of community.

An and Tung (2016) pointed out that "The community plays the subject role of all activities in their locality. The subject role is shown in the fact that community members are active and decision-makers of activities are members of the community". Then, the community is accountable for its results. Hence, the roles of the community in the development of traffic infrastructure are classified as follows:

- Activity participating role – the most important role in community activities: Participation in the development of rural infrastructure can be divided into specific stages such as identifying needs and planning, estimating and defining mechanisms, contributing material resources, directing construction, participating in inspection.
- Resource contribution: labor, intellectual, material and financial resources. With the policy of the State and the people working together, the State can reduce difficulties in mobilizing resources for the development of rural infrastructure.
- Testing and monitoring of the use of the community in the development of rural infrastructure: the dual roles of both a user and an investor/supervisor can enhance the responsibility of the residential community and the participants in the inspection and supervision of the development of rural infrastructure.
- Management and maintenance over a long time of use of public infrastructure.
- Beneficiaries and effective evaluation: the community in the role of users/beneficiaries has an objective assessment during the development of rural infrastructure.

The participants in the community serves as in the following forms:

- The community of government officials (provincial, district, commune and village levels), whose the role is to orient the development process of rural infrastructure and propose policy mechanisms and guidelines for community organization and participation.
- The mass community (farmer's union, women's union, youth union, veteran's association), whose the role is to participate in the supervision, contribute labor resources and especially propagate and mobilize the development.
- The business community (economic organizations, cooperatives, enterprises), whose primary role is to contribute intellectually, technical and technologically and professionally, to invest and or to contribute material resources.
- The people community (local people or expatriates) whom should be considered as the central and active role in rural infrastructure development. Their main role is to participate in all contents of the development, hence considered an essential role. The role of the government is to organize, guide and coordinate all activities, recognize, encourage and strengthen the role of participating communities, mass organizations and businesses.

Different Types of Community Participation in Rural Transport Infrastructure Development

Tao (2008) listed the types of community participation in rural transport infrastructure development: planning for the participation, contributing labor and financing, monitoring, and maintaining the infrastructure. Another classification pointed out the types of community participation: planning for the participation, training the participants, monitoring and maintaining the infrastructure (Quang Ninh DoT 2019). Synthesizing these views, the authors believe that the types of community participation should include:

- Community participation in planning and identifying demand to upgrade, repair, or build new infrastructure. These tasks are very important, but many local governments overlooked them, leading to resentment from the community, making the community uncooperate and refuse to participate. The community has the right to participate in this process, because decisions in this process will affect their lives.
- Community participation in estimating and participation policy: to estimate and develop policies to participate in development of rural infrastructure (e.g., work estimation, compensation and clearance and obtain the right-of-way). Community participation increases the quality of those tasks since working together improves mutual trust and the ability to solve common problems. Community participation at this stage ensures better project outcomes because the community knows the needs, capabilities and resources to which it will commit to contribute.
- Community participation in contributing resources: the community can contribute financial, materials, lands, and human and intellectual resources to the development of rural infrastructure. Resource contribution is a practical act of participation. Depending on the policy, method of mobilization and the purpose and criteria of the project owners, the contribution may achieve results with different levels and forms.
- Community participation in the construction of the infrastructure: the community can directly do part or all of the construction of the infrastructure. Depending on the size of the projects and the policy to attract participation, it is possible to organize the participation in the form of "the state and the people working together" or the state implements the investment project while the community contributes labor. Communities can organize themselves as families or groups of households to do small works such as alleys, bridges, and culverts.

- Community participation in inspection and supervision: inspecting and supervising the process of organizing and implementing the construction of rural infrastructure. Here, the community aims to improve and ensure the ability to execute the projects on schedule, content, original plan, rational and effective use of resources and investment capital according to the project estimates. These tasks are usually delegated to people or groups of people with experience and expertise in the community. Community participation in supervision may reduce waste, loss and corruption in the process of using capital.
- Community participation in the management and maintenance: the tasks of management and maintenance of the works to make the works better protected. Rural infrastructure buildings have a longer lifecycle and less degradation when the community voluntarily protects them in combination with the process of using, directly participating in periodic maintenance and repair, and complying with the regulations to improve sustainable efficiency.
- Community participation in the usage and evaluation of effectiveness: development of rural infrastructure serves both the needs of local socio-economic development and the travel needs of all types of communities (human communities and business community); it is necessary to take into account the fairness in the distribution of benefits as well as obligations with different beneficiaries. There may be a fee for use and maintenance for groups of beneficiaries that do not contribute, but no fees for groups of beneficiaries that do contribute. The evaluation of the effectiveness of the community ensures objectivity and transparency in the development of rural infrastructure.

THE MANAGEMENT OF COMMUNITY PARTICIPATION IN RURAL INFRASTRUCTURE DEVELOPMENT AND ITS INFLUENTIAL FACTORS

The Management of Community Participation in Rural Transport Infrastructure Development

According to Edelenbos and Klijn (2005), the main motives for involving stakeholders in interactive decision-making are to reduce the veto power of different social actors by allowing these stakeholders to participate in the decision-making processes. These decision-making processes fall into six categories - which are compared in three aspects: the nature and organization involved, the manner in which the process is managed (process management), and its relationship to formal democratic institutions. The different forms of decision-making reflect the growing voluntary management interdependence of public actors over private, private, and other public actors.

Interactive governance is described as a way of implementing policies by which the government encourages the participation of citizens, civil society, businesses and other stakeholders in the early stages of the process. Procedures with community participation involved differ from traditional procedures: parties are involved in formulating policy proposals, while in traditional form, community involvement only occurs when the policy proposal has been developed. Interactive decision making is a policy practice, solutions in managing the participation of many stakeholders; though the solutions are often not innovative enough, or there is a large gap between the stakeholders, public officials and people (Edelenbos and Klijn 2005).

Managing community participation in the development of rural infrastructure is a continuous process throughout: identification of needs and planning, estimation and participation policy, construction, inspection and supervision, management and maintenance, and performance evaluation. The influential factors (objective and subjective) have an impact on the community's participation in the development of rural infrastructure, so it is necessary to develop a management method. There exists the request to study and synthesize the contents related to the community participation, the contents and factors affecting the participation of the community, the experiences and solutions to enhance the participation of the community involvement (McLaverty., ed., 2002).

In short, managing community participation in the development of rural infrastructure with diverse content, forms and methods of participation is an indispensable need for participants. Moreover, depending on the specific conditions of each locality, each community, and the ability of community actors to participate, the form of participation in community management may be diverse or limited.

Influential Factors of the Management for Community Participation in Rural Transport Infrastructure Development

Community Management Qualifications and Experience

MoHA (2012) provided a guideline on the recruitment of civil servants at commune, ward and township levels: at least a professional secondary degree or higher; in mountainous, border, island, deep-lying and remote areas, the lowest degree may be reduced by one level compared to regulations. However, commune and village officials are the main force to mobilize and guide the community, so the results of community mobilization depend greatly on the enthusiasm and capacity of the organization to implement.

Mechanism and Model of Community Participation Management

The operating mechanism of the Management Council in public non-business units in the field of transport in general and rural infrastructure is regulated as follows: The Management Board meets periodically at least once every three months and holds extraordinary meetings at the request of the Chairman of the Board. The Management Board works in a collective mode; its meetings are considered valid when at least 2/3 of the members attend (MoT 2018).

Methods and Organization of Community Participation Management

In practice, the organization and management of community participation have not been carried out in a methodical manner, leading to many negative consequences. The results of community participation have not yet ensured publicity and transparency, limiting wide participation and continuous participation of the community. The effectiveness of participation is limited due to the lack of planning for each stage of participation (Hai and Quyet 2015). Community participation management organizational planning can refer to and use existing tools such as the Integrated Rural Approaches (IRAP) as a tool to produce comprehensive plans to improve accessibility situation for a community, area and/or region. IRAP is often used in the context of "normal" development. A stripped-down version of the process can be used in a crisis to quickly identify priorities in a participatory manner. Accessibility actions are developed at the district level, which serves as a central tool to guide immediate investments in infrastructure and will prove useful for long-term reconstruction programs. The investments identified will include rural infrastructure and roads (WB 2019).

Objects and Characteristics of the Community Involved

Rural infrastructure system is associated with all economic and social systems while management participation from the community in development rural infrastructure is a factor promoting socio-economic development; both depend on the level of socio-economic development of the countryside. Rural infrastructure is a complex structural system distributed throughout the territory, in which there are parts with different high and low levels of influence on the socio-economic development of the entire countryside, regions, villages and communes. However, these parts have a close relationship with each other during operation and exploitation (Quyet 2016).

Other Factors That Influence Participation Management

Management model of community participation in development of rural infrastructure features outstanding participation of all members of the village community. It is a collection of all management components for development purposes. This has increased the connection between the representative board and other organizations in a village or in a commune. The village head is the person who is primarily responsible for administration and implementation of all general activities of the village. In addition, the representative board coordinates with associations, mass organizations such as Youth Union, Elderly Association, Farmers' Union, and Women's Union.

The effectiveness of community participation management in rural infrastructure development is accessed regarding to the participation roles of different types of communities in different aspects and stages of rural infrastructure development. Depending on the ability to organize and manage the participation of different types of communities, the role of government participation is important or less important. The enhancement of the participation role of all types of communities can be improved by promoting training, capacity building, organizational skills, and participation management for community organizations. This means that the government can change its own role in managing the participation of community organizations.

Practical Experience For Managing Community Participation in Rural Transport Infrastructure Development

International experience in managing community participation in rural transport infrastructure development as:

- *In Japan*, Since the 70s of the last century, the "one product per village" movement was initially implemented in Oita prefecture (southwestern Japan). Since then, after nearly 30 years of implementation, this movement obtained many important results: the movement "one product per village" had attracted the attention of many other regions and countries. The movement encouraged and utilized local resources, promoted community strength, preserved traditional craft villages and narrowed the gap between urban and rural areas. The success of this movement is arguably an operational organization, model, method of resource management to professional level. But the movement was far from the ability to build a participatory management model and method for the community in the development movement building and stable future development (Quyet 2016).

- *In Korea*, the Korean Rural Innovation Movement (referred to as New Village), though launched more than 40 years ago, still motivates the Korean people. The movement "New villages" focuses on the following ten ways of implementation: Expand and renew roads in the village; Clean villages; Construct a common laundry; Dig common wells; Upgrade thatched to tile/cement roof; Upgrade earth wall to brick/cement; Repair Bridge; Repair river dam system and construct northern composting point. After 10 years of implementation, the movement contributed a thorough change in the face of rural Korea; moreover because this movement built and managed a massive rural transport infrastructure system, and diverse community participation also operate under a unified and management approach (Quyet 2016).
- *In Malaysia*, the government has paid attention to reduce poverty by improving rural infrastructure development. Malaysia is determined to provide a broader and more effective rural transport infrastructure system through the mobilization of more resources. There are policies to attract investment capital of the private sector through Build-Operate-Transfer, Build-Transfer-Operate, Build-Transfer investment forms, with support for science and technology, human resources, and calling for the participation of the people. As a result, the speed of rural transport infrastructure development is relatively fast, immediately meeting the needs of rural development. However, the model is yet to maximize the participation of the community. For instance, people are not allowed to participate in activities such as contributing ideas, and preserving and maintaining assets. Private participation in

investment has been distorted, especially toll collection tasks (Quyet 2016).

- *In Thailand*, the Thai Government issued the government's investment policy and mobilization of participation with a special focus on developing inter-village roads. The Third 5-year plan (1972-1976) prioritized key routes linking production zones with processing and consumption markets, ensuring that the distance from villages to any car route does not exceed 5 km to enhance the inter-village traffic network with a provincial, district, and commune administrative borders and that investment in roads is a priority of the government to serve the interests of the people. Thailand focused on the subjects involved in the management of government agencies and the business community to develop production such as agricultural production enterprises and agricultural cooperatives; households were not really encouraged to participate, so the effectiveness and results of participation were still limited (Tao 2008).

Experience of Some Localities in Vietnam on Managing Community Participation in Rural Transport Infrastructure Development

Along with the economic opening, the diversification of forms of participation of people and communities in rural transport infrastructure development is indispensable and evident through the contribution of capital mobilized from the people (Quyet 2016). The government in Phu Tho province based its policy on the specific criteria of the New Rural Construction Program, maximizing the role of the people in rural transport development with the motto that rural transport was "for the people, by the people, and by the people themselves." All work is done openly and democratically. From the very beginning, it was necessary to maximize the role of the community, mobilize all economic sectors to participate, especially the community of people in the development of rural infrastructure. The government has developed a mechanism for mobilizing participants to contribute flexible resources, so the mobilization of economic sectors to participate is better, and the participation results are higher. In addition, in the propaganda, advocacy and organization and management of all rural infrastructure development activities were based on the people of Phu Tho province. It is considered a successful solution to involve the community (ADB 2012).

Nam Dinh was one of the first two provinces in the country to complete the task of building a new countryside. Rural infrastructure is invested and upgraded quite synchronously, creating favorable conditions for socio-economic development, improving living and working conditions of rural people. Investment resources from the state budget are only a part (about 26%), the rest is mainly mobilized from the people and socialized. The role of the people in the construction of a new rural area is promoted, taking the people as the subject, relying on the people's strength; the outstanding results in the construction of new rural areas of Nam Dinh province all have great contributions from the people. People participate in the construction of a new rural area mainly through five forms: (1) participating in the planning and project of building a new countryside at the commune level; (2) participating in the selection of jobs that need to be done first and done later, which are both practical for the people and suitable to the local capacity and conditions; (3) deciding on the level of contribution in the construction of public works of the village and commune; (4) appointing representatives to participate in the management and supervision of construction works; (5) organizing the management, operation and maintenance of the works after completion (Phong 2019).

Binh Duong, typically Bau Bang district, has developed a plan on the development of rural transport routes in the area. The district has hardened rural traffic routes with the motto "The State and the people working together". In the period of 2014 – 2019, the movement to build rural roads in the area was practical and effective, pleasing to the people in the propaganda and mobilizing people to join hands with the State to implement the plan.

To successfully achieve the set goals, it is necessary for the government to develop a management model for community participation in rural infrastructure development, in which it guides the review and perfect appropriate mechanisms, policies and criteria for building new rural areas with the characteristics of each region.

THE MANAGEMENT OF COMMUNITY PARTICIPATION IN THE DEVELOPMENT OF RURAL COMMUNITY INFRASTRUCTURE IN MEKONG RIVER DELTA, VIETNAM

Introduction to the Mekong Delta, Vietnam

The Mekong Delta is a part of the Mekong River area with an area of 39,734 km². This area is located adjacent to the Southeast region, the North borders Cambodia, the Southwest is the Gulf of Thailand, the Southeast is the East Sea. The poles of the delta on the mainland, the western pole at 106°26´ (MyDuc commune, HaTien town, KienGiang province), the eastern pole at 106°48´ (TanDien commune, GoCongDong district, TienGiang province), the North pole is at 11°1´ (LocGiang Commune, DucHoa District, LongAn Province), the South Pole is at 8°33´ (DatMui District, NgocHien District, CaMau Province). The offshore islands of Vietnam such as PhuQuoc island, ThoChu archipelago, Khoai island are also in this area.

The agricultural production value of this region leads the country, accounting for 33.2% of the total value of national agricultural production. The industrial production value of the whole region accounts for about 9.23% of the total industrial production value of the country, ranking after the Southeast Region and the Red River Delta.

Currently, all provinces and cities in the Mekong Delta have approved provincial planning tasks. Therefore, the collection of opinions on the regional planning also aims to ensure the linkage, synchronization, and consistency among the master plans in the context that the planning's are being developed at the same time. The rural infrastructure system in this area accounts for about 13% of the country and is considered a place with slow development of transport infrastructure, not commensurate with its potential and has not aroused the motivation to create economic breakthroughs and ensure social security. This is also a topic of interest to the Congress (MoC 2021).

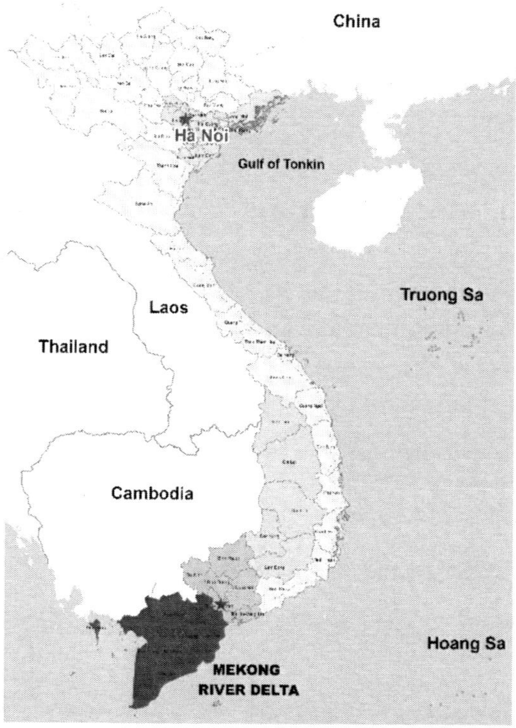

The Mekong River Delta consists of 13 central administrative units: 12 provinces of LongAn, TienGiang, DongThap, VinhLong, TraVinh, HauGiang, SocTrang, Ben Tre, AnGiang province, KienGiang, BacLieu, CaMau and CanTho city. It is one of the six economic regions of Vietnam; the region's strengthest agricultural production.

Figure 2. Maps of Vietnam and the Mekong River Delta's Provinces.

Current Situation of Community Participation Management in Rural Transport Infrastructure Development in the Mekong Delta, Vietnam

Community Management Qualifications in Rural Transport Infrastructure Development

The team of grassroots ministries are the ones who directly develop plans, set measures, and organize the implementation of community participation management activities in rural infrastructure development, thus improving the capacity of all stakeholders. For this team, it is very important to realize the management of community participation in the development of rural infrastructure. Practice shows that most grassroots officials are weak in terms of professional knowledge and skills in the field of community participation management (BacLieu PPC 2020 and DongThap PPC 2020).

Mechanism and Management Model in Rural Transport Infrastructure Development

The policy mechanism and model of management of community participation in the construction and development of rural infrastructure show many shortcomings, lack of democracy and publicity. The management of investment capital is limited; the management capacity has not met the requirements of continuous development, resulting in many mistakes in the management of construction activities in the localities. This situation caused suspicion in the community; there are even cases where complaints and lawsuits arise, causing community disunity (Su 2019).

Characteristics of Rural Transport Infrastructure with Participatory Management from the Community

The participation of community in the construction of the rural infrastructure project brought about a great change in the development governance structure, combined with the change of landscape and had a positive impact on the activities of the rural population because the community is the direct implementers and beneficiaries. These characteristics come from many different aspects of the location of the rural transport system, with high investment, long service life, paving the way for economic and social development. Rural infrastructure, of the region or of the village or commune, needs to be formed and developed one step ahead and suitable for socio-economic activities. Community members participate early from the planning stages of rural infrastructure project for socio-economic development to decide on the construction of rural infrastructure project in terms of scale and quality, reflecting the economic development orientation, society and create a material premise for the process of socio-economic development in rural areas (PMoVN 2012).

Methods and Management Organization in Rural Transport Infrastructure Development

The management of exploitation and the management of the participation of communities in rural infrastructure development such as rural roads will not be planned separately from other necessary infrastructure. The result from planning to participation management is a list of communities that will participate, and the representative organization that manages that participation in the model. Priority rural infrastructure projects for construction such as rural roads and some other infrastructure projects serving the life of rural communities. The key point is that the community participation management model has not been paid attention to, all of which are managed and operated by the district/district/ward/commune/political organizations/organizations and do not have an appropriate model. Though proven appropriate for each field of participation management, methods and organizations are still spontaneous with many negatives and lack of transparency (TraVinh PPC 2020).

Objects and Characteristics of the Community in Rural Transport Infrastructure Development

The object of management of community participation in rural infrastructure development in the Mekong Delta is characterized by the participation of all members of the village community. It is a collection of all components for development purposes. This has increased the connection between the representative board and other organizations in the village and in the commune.

Model of Community Participation Management in Rural Transport Infrastructure Development

The interaction between associations and organizations and mass organizations in the village has created a unified whole, under the management and necessary support of the commune People's Committee,

Fatherland Front Committee, mass political organizations which act as expert advisory bodies for management and methodology development. Thus, the establishment of the representative board has attracted the participation of all households, from rich households, well-off households to poor households, and from men to women. The members of the representative board are all members of the village associations and organizations. This has proved that equality among all people is the same, regardless of the income levels and regardless of genders. Therefore, the representative board plays a very important role in self-mobilizing contributed capital, labor and property of people in the village or their children working in other localities. The mass organizations promote the role of the masses so that all households enthusiastically participate in the local rural development movement (SocTrang PPC 2020).

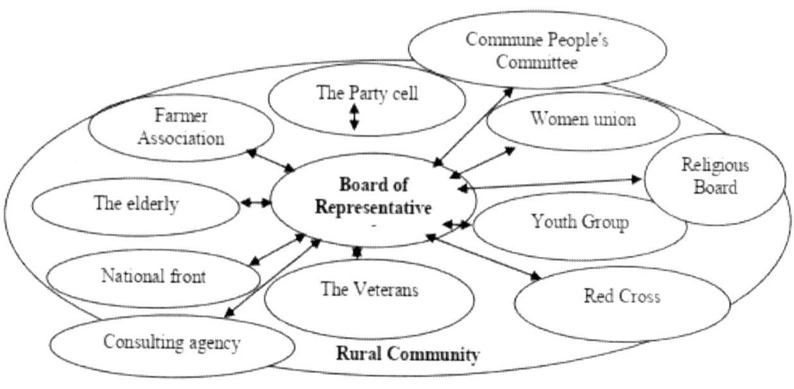

Figure 3. Relationship between the Board of Representatives and community organizations.

Effective Management of Community Participation in Rural Transport Infrastructure Development

The effectiveness of community participation management in rural infrastructure development is different when measured by the participation roles of different types of communities: the government is important to the inclusiveness such as identifying planning needs, budgeting, and policy for participation. Depending on the ability to organize and manage the participation of different types of communities, the role of government participation is important or less important. The enhancement of the participation role of all types of communities can be improved by promoting training, capacity building, organizational skills, and participation management for community organizations. This means that the government can change its own role in managing the participation of community organizations as follows:

- Management of community participation in the commenting stage: determining the need for rural infrastructure development planning is necessary to prioritize the development of the most urgent types of rural infrastructure in the district. The content of the community participation in this phase includes: comments, request information on quality rural infrastructure, status of types and development needs of rural infrastructure. The results of the community participation in contributing ideas on the development planning needs of rural infrastructure are assessed by the ratio of the number of members of each type of community participating to contribute ideas to the number of people invited to participate, by type rural infrastructure, and the extent of using comments for this content in practice in Bac Lieu province (BacLieu PPC 2020).
- Management of community participation during the development and estimating stage and policy for participation in rural infrastructure development include the following participation contents: Contribute ideas for the estimation of agricultural infrastructure works villages regarding construction plan, implementation schedule, total value of works and mobilized capital sources; Develop a policy of community participation such as regulations on the form of contribution (money, labor, materials or land), the level of contribution of each type of household (poor/well-off, policy households) and identify the rights and obligations of the community to participate.
- Management of community participation during the construction phase: government officials often play the role of organizing, managing and supervising construction, with rural transport investment projects without authorized government as the investor.

There are also cases where the government, mass organizations and community representatives discuss with professional and technical units (units hired for construction) and supervise them. The government can also organize, train and provide legal support to the community in terms of organization and expertise in case the community organizes construction by itself.

- Management of the community participation during the testing phase: monitoring to ensure the rural infrastructure is closest to the needs of the community. The effectiveness of this task depends on community outreach and training. In this study, all 120 votes/100% of the surveyed communities, all serveyees said that they would participate if they were required and received professional training and supervision, and the percentage of communities that continued to participate was higher than that required. Community supervision on the development process of rural infrastructure mainly focuses on collection and expenditure of contributions, construction, estimation, and finalization of investment projects.
- Management of the community participation during the operation and maintenance: the local government is the investor or is authorized to be the investor; so when the construction is finished, the asset will be handed over to localities according to management level. Investment projects carried out by community organizations themselves, the government then played a supporting role in guiding legal procedures and guiding the operation, exploitation, and maintenance, maybe in the form of "self-governing works". The business community participation in this content is primarily due to a regulatory requirement for the management activities that the business communities themselves are engaged in construction, repair, or maintenance. Voluntary participation is only done because the business communities operating in the area that the damaged rural infrastructure are required to repair. In contrast, the community participating in this stage is mainly voluntary in the spirit of movement activities, so there is a higher level of participation results (DongThap DoARD 2020).

SOLUTIONS TO PROMOTE MANAGEMENT COMMUNITY PARTICIPATION IN RURAL INFRASTRUCTURE DEVELOPMENT IN COMMUNITY PROVINCES IN THE MEKONG RIVER, VIETNAM

The Ministry of Transport issued Decision 622/QD-BGTVT to approve the investment project to build residential bridges and manage local road assets (LRAMP - Local Road Assets Management Project) funded by the World Bank. In the Decision, the Ministry of Transport encourages localities in the Mekong Delta to use traditional solutions. Specifically, traffic planning must be consistent with other specialized plans. In addition, it is necessary to develop mechanisms and policies to encourage all economic sectors, small and private contractors to participate in construction, operation and maintenance of rural transport and provision of regional services. Stemming from the research results and the actual situation, the authors propose solutions to strengthen the management of community participation in the development of rural transport infrastructure in the Mekong river Delta, as follows:

Building Models and Coordination Mechanisms for Community Representatives to Manage Community Participation in Rural Transport Infrastructure Development

The results of the analysis of the group of factors "Participation management mechanisms and models" showed that, for effective participation management activities, it is necessary to have a community management model suitable for each locality and mechanism. Recognition of contributions to participation management is necessary since community participation in localities depends on economic conditions, socio-cultural characteristics and perceptions of each local community, participation models and mechanisms vary widely. Furthermore, policies for community management need to be transparent because this requirement is crucial in effective participation management.

Transparency is also a factor in maintaining and developing sustainable participation, encouraging community participation to contribute. Therefore, it is necessary to have a mechanism for public management of the community and ensure the grassroots democratic regulation throughout the management process along with the participation of the community.

From the practical research, the authors propose a model to manage the participation of community representatives in the development of rural infrastructure to overcome the limitations of the participants and enhance the participation of the community. The mechanism should provide flexibility in the process of participating in management, thereby improves the management efficiency of community representatives (Figure 4).

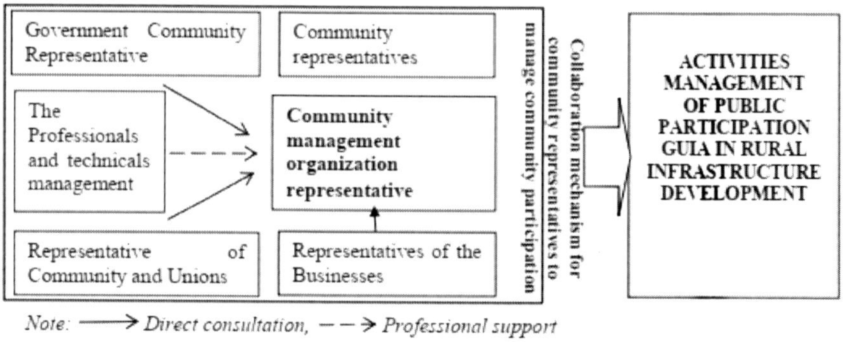

Figure 4. Models and coordination mechanisms for community representatives to manage community participation in rural transport infrastructure development.

1. Community of government officials: participate in orientation, advice, training and professional guidance for the participation process of all communities. Government officials need to rely on the guidelines and mechanisms of the State to build a governance mechanism and encourage the participation of the community in accordance with the conditions and characteristics of each locality, build and maintain democratic mechanism for community participation, promoting the leading and leading role.
2. Community: community contributes financial, human labor, and intellectual resources. Representing the community can be a member or a group of members or an entire community

organization participating in any form of management. The community is both a development participant and a beneficiary, so it needs to participate in all the contents of planning, construction, inspection and supervision of rural infrastructure construction activities, especially construction projects. These contents should ensure the most democracy and publicity, such as determining the planning needs; budgeting and participation policies; and contributing material resources and monitoring. Therefore, with the role of participation throughout the process, the community needs to participate in training to improve the effectiveness of participation.

3. The business community: this is the participant who provides the main task of participating in contributing resources with the mechanism of calling for investment, combined with calling for participation as a subject to participate in construction management and benefit. This participant needs to be encouraged to participate in a variety of forms and methods in the process of building rural infrastructure. In fact, the form of management participation in the form of public-private partnership (PPP) for the development of rural infrastructure has been adopted by some Vietnamese localities, such as BinhDuong province, HoChiMinh city and other developing countries.
4. Community and mass organizations: they have an important role in managing and organizing community activities as well as serving as a core part of the representative of the community organization. Supports to community management by local authorities are mainly propaganda, participation mobilization and representation, supervision and management of rural infrastructure construction through movement activities to attract members. The mass community needs to associate its activities with the contents of community participation management in rural infrastructure development and the regular participation management performance evaluation criteria. The contents of the local government's participation in management can be ideas on

the community participation management mechanism, labor, supervision, operation, and maintenance.

5. Representatives of the organization of the community management board: they are the personnel's nominated from the participating communities, whose roles and responsibilities are to directly manage all participation activities, build the participation mechanism, implement and monitor all community participation activities. This organization also represents the community in planning, organizing, implementing and taking responsibility for the results and effectiveness of local rural infrastructure development activities, and is responsible to the community for the results and effectiveness of community participation. Since the representatives of the community management board are representatives of different communities participating in the management, the government needs to develop a coordination mechanism for the community representatives to manage compliance and ensure the organization of management activities.

Coordination mechanism for community representatives to manage participation: it depends on the application of the community participation management model in rural infrastructure development, which is suitable for each locality and the characteristics of the organization. Specifically with the model of the state as the investor, the representative of the community management organization board will be supervised by the government to manage and organize community participation activities, or the self-development model of the People's Community. The representatives of the community organization are elected by the people, then they begin managing and organizing the participation.

6. Specialized technical units: specialized technical units are often hired by the government or representatives of community organizations to estimate, construct, construct, and supervise the contents of the project. However, technical experts might face difficulties handling the process of building and developing rural

infrastructure in the context existing participation from community. Technical experts therefore need to get familiar with lessons learned provided by the government before they can effectively get involved in the process.

In summary, the model and the coordination mechanism to represent the community to manage the community participation in the development of rural infrastructure will help to manage and coordinate the participation of community management actors, ensuring flexibility in participating management activities under the coordination mechanism. Proposed model will support community participation management to overcome the lack of connection of stakeholders in management and community participation in rural infrastructure development.

Improve Management Qualifications for Community Representatives to Manage Participation in Rural Transport Infrastructure Development

Practical research shows that most grassroots officials are still weak in terms of professional knowledge and skills in the field of community participation management. The situation of degeneration and corruption in cadres and party members is still complicated and slow to be overcome (BacLieu PPC 2017). Attention should be paid to professional training in public investment management, strengthening advocacy skills, and organizational skills for all officials directly involved in the direction and implementation of public investment activities (DongThap PPC 2019).

Develop Methods and Organize the Management of Community Participation in Rural Transport Infrastructure Development

Community participation management organizational planning refers to and utilize existing tools such as the Integrated Rural Approaches

(IRAP) to produce comprehensive plans to improve accessibility situation for a community. IRAP is often used in the context of "normal" development. A stripped-down version of the process can be used in a crisis to quickly identify priorities in a participatory manner. Accessibility actions will be developed at the district level, which will be a central tool to guide immediate investments in infrastructure and will also be useful for long-term reconstruction programs. The investments identified will include rural infrastructure and roads.

According to Vinh and Vinh (2012), types of community management organizations include: government agencies, businesses, banks, schools, and non-governmental organizations. These organizations are formally organized and systematically manage the activities of society. Community development is the local community actively mobilizing internal resources before seeking external support. People in the community (not government agencies or NGOs) are at the heart of initiating, designing, and implementing participatory development activities. There are two means by which to build and manage community:

1. Voluntary groups/organizations with an informal, rotating organizational structure. The form of participation in management is decided by the members on the direction of action and elects the executive and management board. The form of participation is voluntary, meaning no salary, so the operation is based on volunteering participation, information exchange through sharing stories, practical experiences, method-based community management intermediary, and guide. Decentralization method is based on consensus and equality.
2. Government/formal organizations, formal organizational structure, hierarchical system, with participation forms organized to perform different tasks. The management here is appointed, distributed public employees of executive boards, managers. Therefore they are paid with salary, part-time allowance, or voluntary part-time without salary.

Regularly Train Community Representatives to Manage and Advocate for the Community and to Participate in the Management and Contribution For Rural Transport Infrastructure Development

Implementing Resolution 120 of the Government in the period of 2021-2025, the Ministry of Agriculture and Rural Development prioritizes allocating and mobilizing resources from the community for investment projects rural infrastructure serving agricultural production. The results are expected to stabilize the lives of people in the Mekong River Delta while recognizing that one of the limitations is that the level and capacity of management are still weak, yet to meet the complexity of the system.

Therefore, it is necessary to train community management for organizations, including knowledge related to mechanisms and policies, experience in community organization and management in each development content rural infrastructure, from identifying needs bridge; budgeting and participation policy; managing the participation and contribution of resources; construction management; supervision and acceptance; management and maintenance; rural infrastructure effectiveness assessment and community participation. It is necessary to train the community about specific mechanisms and policies related to the community's activities and forms of participation, such as: participating in providing information, contributing ideas, contributing financial resources, and human resources. From there, the people who need to participate in the training are representatives of community management organizations, professional boards, full-time government representatives, mass organizations and related professional units. In each different participation activity, the program and content to be trained are different, suitable for the representative of the community organization in charge of grasping and implementing more effectively.

Moreover, the factors (i) the economic conditions of the participating communities are difficult, (ii) the characteristics of rural communities are complex, (iii) the sense of rural community, which is cohesive reflecting the situation of rural communities, is very complicated. However, a high sense of cohesion should be noted in the process of community participation management; this is considered a strong point of the community in which activities are carried out. Therefore, the government needs to propagate and mobilize the participation of the community according to the subject of the community and the characteristics of the local area, in combination with the target program of socio-economic development and community activities. The local community is an effective solution for propaganda to call for participation. The practical programs for the Mekong Delta provinces are advocacy for management and community participation in the development of rural infrastructure combined with the New Rural Program with the criteria for developing road infrastructure and propaganda calling for the participation of the community in development rural infrastructure in combination with the Program of Agriculture - Farmers - Rural areas.

In short, one of the communication solutions to mobilize community participation in the development of rural infrastructure is based on the combination of communication about the goals of socio-economic development and development of rural infrastructure.

LIMITATIONS AND FUTURE RESEARCH

One of the limitations of this study comes from the structure of data analyzed. The primary source of data included 120 surveys, which could be extended in quantity to provide a more robust observation. The data were collected mostly in DongNai and surrounding provinces, since this area can be considered a good representative of the Mekong delta. However, the data can be extended geographically for the researchers to make a more solid representative of community participation.

Many new policies have been issued in the period of 2016 – 2021, yielding consequences that have been reported. However, to fully assess the effectiveness of those policies, the government and others should encourage further research in the next few years. Therefore, the authors plan to conduct a few studies: (i) expand the scope of the research to cover the whole area of the Mekong delta and possibly other areas in Vietnam, (ii) conduct analysis on the effectiveness of the four proposed solutions, especially building models and mechanisms, devising methods, and training the community about community participation in rural infrastructure.

Conclusions and Recommendations

Conclusions

Managing community participation in rural infrastructure development is quite complicated because of (i) the specific nature of the object and (ii) the process is involved with many stakeholders with widely ranging capacity, qualifications and experience in community management. Therefore, it is necessary to develop a model and a participatory management mechanism that is flexible, suitable to the conditions of each locality, combined with the development and encouragement of management participation in parallel with community contributions. Successful mechanisms will contribute to the improvement of the efficiency of community participation management in rural infrastructure development in the Mekong Delta provinces of Vietnam.

Implications can be driven from the four solutions that the authors proposed. These implications provide commercial opportunities for the consultant's sector to serve on-contract consultings to the government in studies about building model and mechanisms of the community participation in project development. The private sector can provide training courses for the public so that the public understand more about infrastructure development with community participation.

There are many factors that affect community participation, such as community management qualifications and experience, participatory community management mechanism and model, methods, organizations to manage community participation and objects and characteristics of the community involved in the management of community participation in development rural infrastructure. Therefore, it is necessary to flexibly combine solutions to manage community participation in rural infrastructure development according to the characteristics of each locality.

Recommendations

For the State

First, the State needs to supplement and perfect mechanisms and policies for flexible participation management for community participation management activities in rural infrastructure development. With the principle of harmonization of interests with orientation, the state only provides legal support as the main, maximum mandate for the community to actively participate and decide on the content, form, and level of participation in the spirit of freedom. The State needs to develop policies to encourage localities to expand forms of community participation in management for rural infrastructure development, to mobilize all resources to contribute, participate in management, and promote the community. Secondly, the State needs to regularly summarize practical lessons and experiences in organizing, managing, and mobilizing the participation of the community in the development of rural infrastructure to promptly adjust the main policies, community participation management policy, and the diversification of socio-economic development programs in rural areas. Third, the State should, through training institutions, political organizations, mass organizations, media, and communication tools, disseminate policy topics to mobilize participation and manage participation. Thereby, the State can improve the effectiveness of community participation management and community participation in the development of rural infrastructure.

For Localities

Due to the characteristics of participation, the diversity of participating communities, and forms of participation, localities need to flexibly apply solutions to enhance management of community participation in rural infrastructure development, to ensure feasibility and suitability with the characteristics of each region, each rural area (districts, communes, and village) in the provinces of the Mekong Delta, Vietnam.

ACKNOWLEDGMENT

This work has been sponsored and funded by HoChiMinh City University of Food Industry under Contract No. 69/HĐ-DCT.

REFERENCES

ADB – Asian Development Bank (2020). *Increased participation for development results.* ADB Avenue, Mandaluyong City 1550 Metro Manila, Philippines. Retrieved December 18, 2020, from https://openaccess.adb.org/bitstream/handle/11540/3228/strengthening-participation -development-results-vn.pdf? Sequence = 3.

An, L.V., and Tung, N.D. (2016). *Handbook of Guidelines: Community Development (A document for people involved in community development).* Thanh Nien Publishing House

An, L.V. and Tung, N.D. (2016). *Manual for Community Development - A Document for Community Development Workers.* The document published by the Japan Agency for International Cooperation. Youth Publishing House.

BacLieu PPC - Provincial People's Committee (2017). *Decision No. 44 / QD-UBND dated April 4, 2017, promulgating the Regulations and guiding the implementation of criteria for new rural communes in the period of 2016 - 2020 in the locality.* Bac Lieu Province.

BacLieu PPC - Provincial People's Committee (2020). *Plan No. 75 / KH-UBND, dated June 25, 2020.* Plan on Repairing, renovating, and upgrading rural roads for new rural construction in the form of the State and People working together in Bac Lieu province.

Bellone, Cinzia, and Vasiliki Geropanta. The 'Smart'as a Project for the City Smart Technologies for Territorial Management Planning Strategies. In *International Conference on Intelligent Computing & Optimization*, pp. 66-75. Springer, Cham, 2018.

Bellone, Cinzia, Pietro Ranucci, and Vasiliki Geropanta. The 'Governance' for Smart City Strategies and Territorial Planning. In *International Conference on Intelligent Computing and Optimization*, pp. 76-86. Springer, Cham, 2018.

Clanrence S. (1961). *Scientific, Technical, and Related Societies of the United States*, National Academy of Sciences.

DongThap DoARD - Department of Agriculture and Rural Development (2020). *Proposal No: / TTR-SNN, Regarding the promulgation of the Plan to implement the National Target Program on New Rural Construction in 2021.*

DongThap PPC - Provincial People's Committee (2020). *Implementation Plan of the National Target Program on New Rural Construction in 2021.*

Edelenbos, J. and Klijn, E.H (2005). Managing Stakeholder Involvement in Decision Making: A Comparative Analysis of Six Interactive Processes in the Netherlands. *Journal of Public Administration Research and Theory Advance* Access published July 6, 2020.

Gustav, A., and Thomas, N. (1997). *Principles of rural sociology*, Boston, New York [etc] Ginn and Co. Access time 04/22/2020 from http://www.abebooks.co.uk/book-search/title/principles-of-rural-sociology.

Hai, D.T., and Quyet, N.X. (2015). Evaluate influencing factors and solutions to enhance community participation in rural transport infrastructure development in Dong Nai province. *Journal of Construction Economics*, Number 04-2015, Page 37-46.

Hien, L. (2006). *Experience in mobilizing people's participation in development projects: Lessons learned from the construction of the GUOL house in Thuong Long commune, Nam Dong district, Thua Thien Hue province.* Retrieved on February 4, 2021, from http://www.crdhue.com.vn/modules.php?name=Pages&go=showcat&cid=13.

Hung, N.M. (2011). *People know people discuss, people do, people check - From the policy to the reality in our country today*. Retrieved August 21, 2021 from http://tapchiqptd.vn/zh/an-pham-tap-chi-in/dan-biet-dan-ban-dan-lam-dan-kiem-tra-tu-chu-mong-to-now-in-country-now-today / 3563.html?pageindex=652.

Khanh, N. (2019). *More than 14 billion USD for rural roads. Traffic Newspaper.* Source: https://www.baogiaothong.vn/hon-14-ty-usd-lam-duong-nong-thon-d438368.html. Retrieved at: 09:27 on February 16, 2021.

McLaverty, P., ed. (2002). *Public participation and innovations in community governance*. Aldershot, UK: Ashgate.

MoC - Ministry of Construction (2021). *Early completion of the Master Plan for the Mekong Delta for the period 2021-2030, with a vision to 2050*. Source: https://moc.gov.vn/vn/tin-tuc/ 1184/66553 / The year-to-date-planning-to-land-in-the-country-for-long-time-2021-2030 - tam-nhin-to-year-2050.aspx. Retrieved at: 14:38, March 15, 2021.

MoHA - Ministry of Home Affairs (2012). *Circular No. 06/2012 / TT-BNV dated 30/10/2012 of the Ministry of Home Affairs guiding the recruitment of civil servants at the commune, ward, and town levels.*

MoT - Ministry of Transport (2018). *Circular 05/2018/TT-BGTVT guiding on Management Boards in public non-business units in the transport sector.*

Park, R. (1936). Human Ecology. *American Journal of Sociology* 17(1), 1–15.

Phong, D.H. (2019). *Nam Dinh - A bright spot in building a new countryside. Communist Magazine*. Source: https://tapchicongsan.org.vn/quoc-phong2/-/2018/815665/nam-dinh---diem-sang-trong-xay-dung-nong-thon-moi.aspx. Retrieved at: 10:29, December 22, 2020.

PMoVN - Prime Minister of Vietnam (2012). *Decree No. 11/2012 / QD-TTG Approving the Transport Development Plan for the key economic region of the Mekong River Delta up to 2020 and orientation to 2030.*

QuangNinh DoT - Department of Transport (2019). *Final report on Model of community-based rural transport maintenance in the context of a new rural construction commune*, June 2019.

Quyen, D. (2015). *Deploy the implementation plan of the New Rural Enhancement in the period 2015-2017 in QuangTrung commune*. Retrieved March 16, 2021, from http://thongnhat.dongnai.gov.vn/Pages/newsdetail.aspx?NewsId=2229&CatId=6.

Quyet, N.X. (2016). *Study on community participation in rural transport infrastructure development in DongNai province*. Ph.D. thesis in development economics, Code 62.31.01.05, Vietnam Academy of Agriculture.

SocTrang PPC - Provincial People's Committee (2020). *Plan No. 32 / KH-UBND, dated February 26, 2020. Plan for the Implementation of the National Target Program on New Rural Construction in 2020.*

Su, H. (2019). Bac Lieu Provincial Inspector discovered many mistakes in Vinh Loi district. *News from Vietnam News* Agency, Source: https://baotintuc.vn/phap-luat/thanh-tra-tinh-bac-lieu-phat-hien-nhieu-sai-pham-tai-huyen-vinh-loi-20191016105041380.htm. Retrieved at: 10:50, October 16, 2019.

Tao, T.S. (2008). *Experience in the organization and construction of community-based infrastructure works*. Retrieved August 16, 2020, from http://www.crdhue.com.vn/modules.php?name=Pages&go=page&pid=20&page=3.

Toennies F. (1887). Germainschaft und Gesellschaft [German community and society]. Kessinger Publishing, LLC (Reprinted on September 10, 2010) TraVinh PPC - Provincial People's Committee (2020). *Plan No. 33 / KH-UBND* dated July 13, 2020, of the People's Committee of Tra Vinh province to implement the Directive No. 58-CT / TU dated April 1, 2020, of the Standing Committee of the Provincial Party Committee on continuing to increase. Strengthen the Party's leadership in the implementation of the National Target Program on New Rural Construction in the province.

Vinh, N.D., and Vinh, D.T. (2012). *Training materials: The local people-owned and internal-based approach to community development.* Center for Education Exchange with Vietnam, KienGiang, April 2012.

WB - World Bank (2019). *Speed Variation Analysis - A Case Study for Thailand's Streets and Roads.*

In: Advances of Machine Learning … ISBN: 978-1-68507-211-7
Editors: Pandian Vasant et al.

Chapter 9

WARNING SYSTEM FOR CRACKED PIPES IN AUTONOMOUS VEHICLES

Yair Wiseman*

Computer Science Department, Bar-Ilan University,
Ramat Gan, Israel

ABSTRACT

The looming autonomous vehicles will be driven by the vehicle itself; therefore, the autonomous vehicles will be less inspected because human drivers often listen to the engine and its noises. The passengers in autonomous vehicles are usually bothered with other things and sometimes even sleep, so there will be no effective human inspection on the engine and specifically on the pipe system. This paper suggests an automatic system that warns about cracked pipes before they are out of use.

Keywords: autonomous vehicle, vehicle pipes, embedded real-time system

* Corresponding Author's Email: wiseman@cs.biu.ac.il.

1. Introduction

Vehicles have a number of pipes for several uses like pipes for gas, air, brake fluid, oil and other uses [1]. Any component of vehicles is potentially subject to wear and tear and these pipes are no exception [2]. Worn out pipes may bring about various malfunctions depending on the worn out pipe [3].

In some cases torn pipe even on the whole is not a problem; whereas other torn pipes can be an acute problem. Pipes supplying hot air to the vehicle's carburetor in hot countries is actually quite redundant and if this pipe is torn, the vehicle can continue going; whereas the pipes of the brake fluids are very essential and cannot be relinquished [4].

This paper aims at developing a system for torn pipes detection. Many times, drivers' attention is distracted even in non-autonomous vehicles [5]; however, autonomous vehicles [6] are even less supervised because no driver is needed; therefore, many times no one will pay attention to the vehicle's engine and its noises, because the vehicle user's attention can be paid for other issues and this user can even take a nap [7]. In addition, many autonomous vehicles will be shared by several users [8, 9]. These users will not be au fait with the typical noises and the typical functioning of the vehicle's engine, so they will not be aware of changes in the vehicle's noises; nor will they notice changes in the vehicle's functioning.

Several researches about automatic detection of malfunctioning components in transportation have been proposed during recent years - fuselage damage detection [10, 11], damaged tires detection [12, 13, 14] and SkyTran tracks automatic inspection [15, 16]. This paper aims at proposing an automatic device detecting a nearly torn pipe and warning the vehicle's user that this pipe is almost torn before the pipe is indeed torn, so the user can take care of this torn pipe, while the vehicle still functions.

2. Axis Aligned Bounding Boxes

We employed polygons. Polygons are minimal shapes that are able to approximately create real objects. It is widespread practice to create models of real objects employing plain polygons. Such a polygons model is called Spatial Data Structures [17]. Several approaches with the aim of reducing the number of polygon checks when employing Spatial Data Structures have been suggested [18].

We have employed Spatial Data Structures to simulate vehicle pipes with the intention of locating frail areas that have the potential to crack and spotting the polygons containing these frail areas.

Actually, Spatial Data Structures are the starting point of Space Partitioning [19] and Bounding Volumes [20]. Space Partitioning is a technique of space sub-partitioning into convex shapes. These convex shapes are called "cells." Every cell maintains a list of several objects that it comprises. The algorithm can sift out polygons that are unrelated to a pipe by employing these cells.

Bounding Volume is a similar technique. The algorithm breaks each object into small components. Next, the algorithm finds out the close-fitting bounding volume for each small component. Then, the algorithm detects the components that might contain a crack. It should be noted that in this technique the sifting out is less time-consuming, because the algorithm can detect only the minimal partly cover bounding volumes.

Applications using Bounding Volumes have been researched over the years. Various alternatives of implementation have been proposed: Bounding Spheres [21], K-DOPs - Discrete orientation polytopes [22], OBB - Oriented Bounding Boxes [23], AABB - Axis Aligned Bounding Boxes [24] and Hierarchical Spherical Distance Fields [25].

We have employed the AABB approach which is one of the most widespread approaches. When using AABB, each bounding volume of an object is represented by its minimum and its maximum values [26]. There are pros and cons for employing AABB. The main advantage of AABB is more compactly encompassing the model components which will generally produce fewer checks. In addition, an object split into bounding volumes is typically faster [27]. Initially, the algorithm analyses each basic element that a bounding volume consists of. Then the algorithm projects the element on the axes and, so as to find out the minimum and the maximum values in each axis. This fast procedure is obviously important in real time systems like a device for automatic pipe check in autonomous vehicles.

On the other hand, AABB has a disadvantage. AABB needs more memory space for saving the information. Some years ago, memory space was limited and sometimes the information was even on remote machines [28]; however, nowadays, memory space is much less limited and even everyday computers have an abundant memory space [29, 30], so the limited memory space disadvantage is not critical; therefore, we could chose the AABB approach, because the advantages that have mentioned above. Specifically, the pipe's warning device is a real time system, so the computation time is very crucial. Therefore, we have come to a decision to adopt the AABB approach.

The bounding volumes tree was produced in a recursive method. In each step, the algorithm produces bounding volumes for the remaining triangles. Then, the algorithm splits the triangle set into two sub groups, followed by recursive calls with the aim of processing each of the sub groups in the same way.

We have employed Bounding Volume Hierarchies which are a tree representing an object [31]. In Bounding Volume Hierarchies each sub-tree is rooted by an internal node and represents a segment of the object.

The trees of Bounding Volume Hierarchies have only a single leaf for each basic component, so accordingly the memory size required for each object is linear in the number of the basic components. This feature also has an effect on the checking time which will be quite faster.

3. ASSUMPTIONS

The Bounding Volume Hierarchies construction is time-consuming. This drawback can be critical if the object is malleable and a deformation is repeatedly required. Therefore, in detection of moving vehicles usually different techniques have been chosen [32, 33, 34, 35].

However, pipes are statically secured in the vehicle and twists are hardly ever made, so the Bounding Volume Hierarchies construction is performed on the odd occasion, so this drawback is beside the point of this paper's aim.

The algorithm scans the tree with the aim of finding possible cracks. The algorithm checks the trees from their roots and in a recursive manner it will continue all the way down to the leaves of the trees.

4. IMPLEMENTATION

As was explained above, the system has been implemented in a recursive manner. We have employed triangles as many other implementations do [36]. Firstly, the algorithm sets a bounding volume for the current group of triangles. Next, the algorithm splits this group of triangles into two sub-groups. Lastly, the algorithm calls itself in a recursive manner to process the two latest sub-groups. If a sub-group contains only one triangle, the recursive call to the algorithm will come to an end.

The reason for the triangles split into many sub-groups is the creation of as small as possible bounding volumes so the object model will be as perfect as possible.

For example, four triangles can be split by the algorithm in two different ways shown in Figure 1. The numbers written in a circle within the triangles denote which sub-group contains each triangle after each split. Figure 1 visibly points out that generating bigger bounding volumes can cause more triangle checks as it is shown in Figure 1(b). This property of the hierarchical checking motivated us to implement a better split algorithm making use of better splits as in Figure 1(a), with the aim of minimizing triangle checks [37].

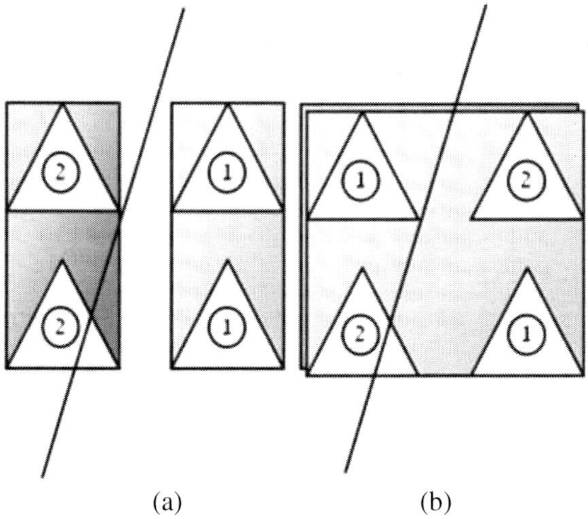

Figure 1. Triangle split (a) Example of split that produces 2 checks; (b) Example of split that produces 4 checks.

Essentially, both the triangle split algorithm and the bounding volume algorithm form the bounding volume tree generation algorithm and control its efficiency. In our implementation, we have employed "Fitting points with Gaussian distribution" [38] as the core of the bounding volume generation algorithm.

The algorithm splits every group of triangles having an analogous bounding volume into two sub-groups. The pseudo code of this algorithm is:

- For each axis of each bounding volume, find a positive direction.
- For each triangle, find the highest valued vertex on the projected axis.
- Sort the triangles' vector by the triangles vertex values.
- For each triangle in the triangles vector:
  - Divide the sum of the projection lengths on the axis by the sum of the original projection lengths of the two sub-groups.
  - Split this triangle with the aim of generating smaller sub-groups.

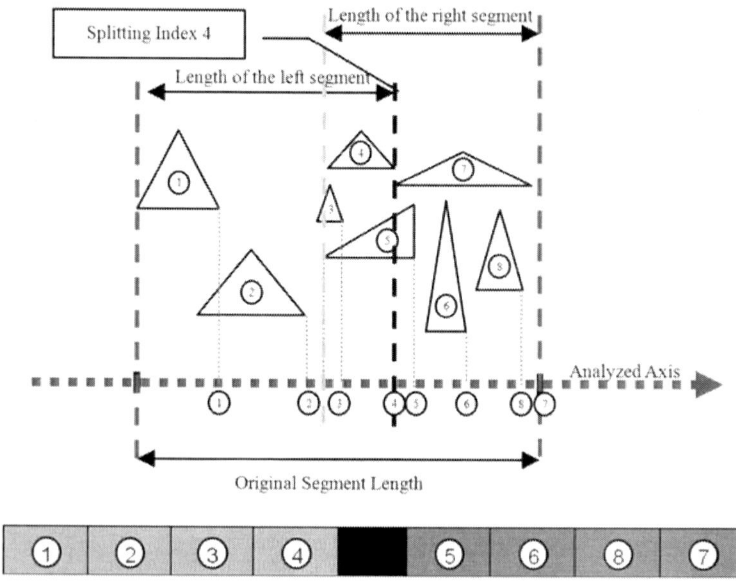

Figure 2. Example of triangles' split on the projecting axis.

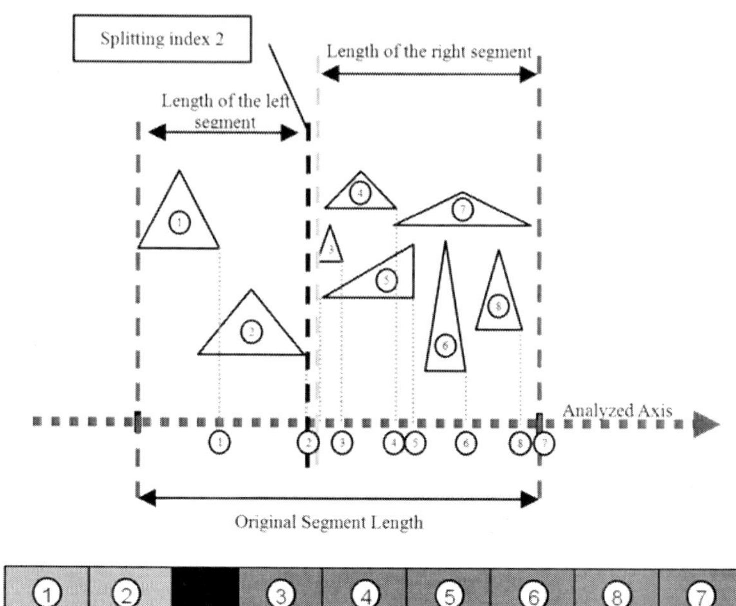

Figure 3. Yet another example of triangles' split on the projecting axis.

The algorithm endeavors to generate as few overlapping sub-groups as possible.

Figure 2 and Figure 3 exemplify two different possible splits. Figure 2 suggests one possible split at triangle index 4. If the algorithm takes this split, a larger overlap between the two generated groups will be created. Figure 3 suggests a better possibility which is a split in triangle index 2. Such a split will generate a much smaller overlap between the two generated groups. Figure 2 and Figure 3 exemplifies when the algorithm should take the best split.

5. Results

The warning system for cracked pipes has been tested on several pipes. The pipe in Figure 4 has been torn in the left side. It is a noticeable rift that can be easily noticed by human eyes; however, an automatic electronic system needs a clear-cut denotation and the triangular split can facilitate such a denotation.

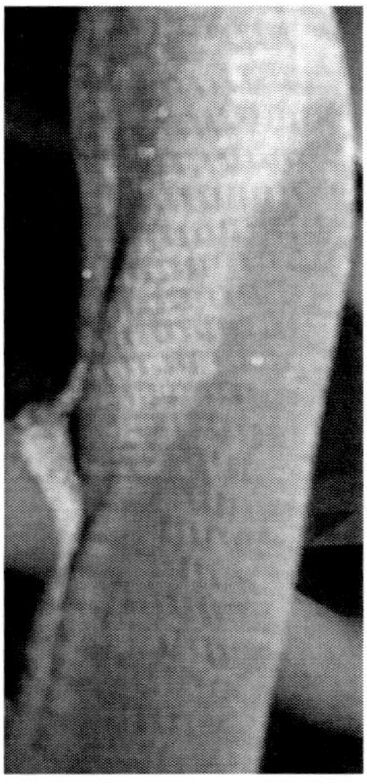

Figure 4. Example of a torn pipe.

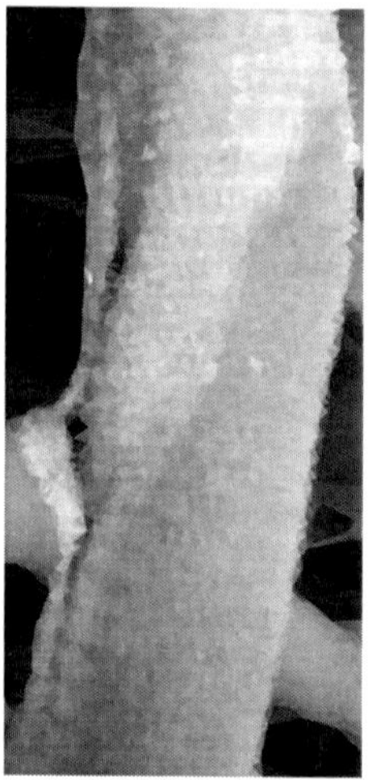

Figure 5. Triangulated drawing of Figure 4.

Figure 5 is a triangulated sketch of Figure 4. Figure 5 can be handled by the warning system for cracked pipes and the considerable difference of the colors can hint on a potential rift.

6. LIMITATIONS

The warning system for cracked pipes is not accurate at all times. False positive cases are a weakness of the warning system for cracked pipes. Figure 6 exemplifies such a false positive case. Some liquids have been dropped on the pipe in Figure 6. Since the color of the original pipe and the color of the dripping are very different, the stains are quite noticeable and the algorithm has considered this pipe as a torn pipe and warned the user.

A triangulated sketch of Figure 6 is shown in Figure 7. Because of the stains, the colors of many triangles have been distorted. These distortions are misrepresented by the warning system for cracked pipes as rifts whereas this pipe is just dirty.

Because of such false positive cases, any warning of this system should be checked by a human in order to verify that the warning is valid and not a mere false positive warning.

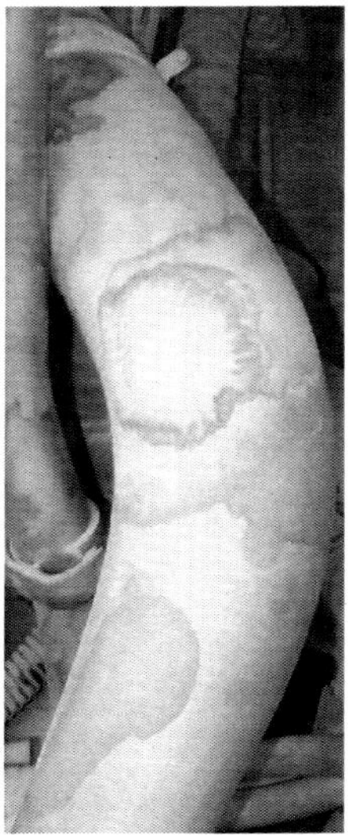

Figure 6. False positive case of an unflawed pipe.

Figure 7. Triangulated drawing of an uncleanly pipe.

7. Future Research Directions

There is an extra execution time when employing OBB (Oriented Bounding Boxes) compared to the AABB technique we have employed. However, the OBB technique can fit the shape of the pipes tighter and do not have to be fully updated if the pipes are repositioned.

In embedded systems it is not common to make use of the most expensive and the fastest CPU; however, if we can afford a much better and faster CPU, choosing the OBB technique might be a more appropriate choice as a bounding volume than the AABB technique. As a future research direction, it would be interesting to check with some experiments, how fast the CPU should be in order to make the OBB technique a reasonable choice.

Conclusion

This paper presents a device that is able to automatically conduct a continuous inspection of potential damaged pipes. This device is especially essential for autonomous vehicles where the passengers are commonly unaware of the vehicle noises, because they do other things. Furthermore, if the vehicles are shared, the users are often changed and so they are not familiar with the vehicle noises.

Continuous inspection along with real time warnings is a common task of computers and autonomous vehicles also adapt this tactic [39, 40, 41, 42]. The suggested device can detect damages before they become tangible and the vehicle is out of use. A timely warning can help to alert the user so as to avoid all of a sudden vehicles coming to a standstill.

Such an intelligent device can be very helpful because detecting damages in pipes before the vehicle stops in nowhere, can avoid annoyance and frustration for the passenger.

REFERENCES

[1] Johnson, T. and Joshi A., (2018), "Review of Vehicle Engine Efficiency and Emissions," *SAE International Journal of Engines,* 11(6), 1307-1330.

[2] Pentapalli, P., Siddiqui, N. A., Mondal, P., and Nandan, A. (2020). "Impact of Speed Breakers on Air Contamination and Mileage of Vehicles," In *Advances in Air Pollution Profiling and Control* (pp. 211-215). Springer, Singapore.

[3] Li, X. D., Yu, Z. W., and Xu, X. L. (2018). "Failure Analysis on High Pressure Fuel Pipes Used in Truck Diesel Engine," *Failure Analysis and Prevention*, 02.

[4] Nadanasabapathy, S., and Kumar, S. R. (2020). "Analysis of fluid behaviour inside the brake lines in non-anti-lock braking system," *Materials Today*, Volume 33, Part 7, pp. 2690-2696.

[5] Alam, L., & Hoque, M. M. (2018). "Vision-Based driver's attention monitoring system for smart vehicles," *International Conference on Intelligent Computing & Optimization*, Springer, pp. 196-209.

[6] Wiseman, Y., (2020), "Autonomous Vehicles," *Encyclopedia of Information Science and Technology*, Fifth Edition, Vol. 1, Chapter 1, pp. 1-11, 2020.

[7] Chui, K. T., Zhao, M., & Gupta, B. B. (2020), "Long Short-Term Memory Networks for Driver Drowsiness and Stress Prediction," *International Conference on Intelligent Computing & Optimization,* Springer, pp. 670-680.

[8] Wiseman, Y., (2017), "Self-Driving Car - A Computer will Park for You," *International Journal of Engineering & Technology for Automobile Security*, Vol. 1, No. 1, pp. 9-16.

[9] Wiseman, Y., (2017), "Remote Parking for Autonomous Vehicles," *International Journal of Hybrid Information Technology*, Vol. 10, No. 1, pp. 313-324.

[10] Wiseman, Y., (2014), "Device for Detection of Fuselage Defective Parts," *Information Journal*, Tokyo, Japan, Vol. 17(9(A)), pp. 4189-4194.

[11] Wiseman, Y., (2013), "Fuselage Damage Locator System," *Advanced Science and Technology Letters*, Vol. 37, pp. 1-4.

[12] Wiseman, Y., (2010), "Take a Picture of Your Tire!" *Proc. IEEE Conference on Vehicular Electronics and Safety* (IEEE ICVES-2010) Qingdao, ShanDong, China, pp. 151-156.

[13] Wiseman, Y., (2013), "The Effectiveness of JPEG Images Produced By a Standard Digital Camera to Detect Damaged Tyres," *World Review of Intermodal Transportation Research*, Vol. 4, No. 1, pp. 23-36.

[14] Wiseman, Y., (2013), "Camera That Takes Pictures of Aircraft and Ground Vehicle Tires Can Save Lives," *Journal of Electronic Imaging*, Vol. 22, No. 4, 041104.

[15] Wiseman, Y., (2017), "Safety Mechanism for SkyTran Tracks," *International Journal of Control and Automation*, Vol. 10, No. 7, pp. 51-60.

[16] Wiseman, Y., (2017), "*Automatic Persistent Inspection of SkyTran Track System*," http://u.cs.biu.ac.il/~wiseman/skytran1.pdf.

[17] Howari, F. M., and Ghrefat, H. (2021). "Geographic information system: spatial data structures, models, and case studies," In *Pollution Assessment for Sustainable Practices in Applied Sciences and Engineering* (pp. 165-198). Butterworth-Heinemann.

[18] Grinberg, I. and Wiseman Y., (2007), "Scalable Parallel Collision Detection Simulation," In *Proceedings of Signal and Image Processing*, Honolulu, Hawaii, pp. 380-385.

[19] Chen, Z., Tagliasacchi, A., and Zhang, H. (2020). "Bsp-net: Generating compact meshes via binary space partitioning," In *Proceedings of the IEEE/CVF Conference on Computer Vision and Pattern Recognition*, pp. 45-54.

[20] Wei, L., Zhou, H. and Nahavandi, S. (2017), "Haptic collision detection on disjoint objects with overlapping and inclusive bounding volumes," *IEEE transactions on haptics*, 11(1), 73-84.

[21] Sung, K. and Smith, G. (2019). "Distances and Bounding Spheres." In *Basic Math for Game Development with Unity 3D*, pp. 87-114, Apress, Berkeley, CA.

[22] Zhou, Y. W., Hu, Z. Z., Lin, J. R., and Zhang, J. P. (2019). "A review on 3D spatial data analytics for building information models." *Archives of Computational Methods in Engineering*, 1-15.

[23] Pijnacker Hordijk, L. (2020). "From Axis-aligned to Oriented Bounding Boxes: An optimization method to reconstruct Oriented Bounding Boxes from Axis-aligned Bounding Boxes of vehicles in low-altitude aerial imagery," *Delft University of Technology.*

[24] Muntoni, A., Livesu, M., Scateni, R., Sheffer, A., and Panozzo, D. (2018). "Axis-aligned height-field block decomposition of 3D shapes. *ACM Transactions on Graphics (TOG)*," 37(5), 1-15.

[25] Funfzig, C., Ullrich, T., and Fellner, D. W. (2006). "Hierarchical spherical distance fields for collision detection. *IEEE Computer Graphics and Applications,*" 26(1), 64-74.

[26] Grinberg I. and Wiseman Y., (2013), "Scalable Parallel Simulator for Vehicular Collision Detection," *International Journal of Vehicle Systems Modelling and Testing, Inderscience Publication*, Vol. 8, No. 2, pp. 119-144.

[27] Melero, F. J., Aguilera, Á., and Feito, F. R. (2019). "Fast collision detection between high resolution polygonal models," *Computers & Graphics*, 83, 97-106.

[28] Wiseman, Y., K. Schwan and P. Widener, (2004), "Efficient End to End Data Exchange Using Configurable Compression," *Proceedings of The 24th IEEE Conference on Distributed Computing Systems (ICDCS 2004)*, Tokyo, Japan, pp. 228-235.

[29] Weisberg, P. and Wiseman Y., (2009), "Using 4KB Page Size for Virtual Memory is Obsolete," *Proc. IEEE Conference on Information Reuse and Integration (IEEE IRI-2009)*, Las Vegas, Nevada, pp. 262-265.

[30] Weisberg, P. and Wiseman Y., (2015), "Virtual Memory Systems Should Use Larger Pages rather than the Traditional 4KB Pages," *International Journal of Hybrid Information Technology*, Vol. 8(8), pp. 57-68.

[31] Jaillet, F., Zachmann, G., Erleben, K., & Andrews, S. (2018). "SIMDop: SIMD Optimized Bounding Volume Hierarchies for

Collision Detection," In *Workshop on Virtual Reality Interaction and Physical Simulation (VRIPHYS-2018)*, pp. 1-3.

[32] Wiseman, Y., "Real-Time Monitoring of Traffic Congestions," *IEEE International Conference on Electro Information Technology (EIT 2017)*, Lincoln, Nebraska, USA, pp. 501-505, (2017).

[33] Wiseman, Y., "Tool for Online Observing of Traffic Congestions," *International Journal of Control and Automation*, Vol. 10, No. 6, pp. 27-34, (2017).

[34] Wiseman, Y., "Computerized Traffic Congestion Detection System," *International Journal of Transportation and Logistics Management*, Vol. 1, No. 1, pp. 1-8, (2017).

[35] Saha, R., Debi, T., & Arefin, M. S. (2020), "Developing a Framework for Vehicle Detection, Tracking and Classification in Traffic Video Surveillance," *International Conference on Intelligent Computing & Optimization*, Springer, pp. 326-341.

[36] Li, Y., Shellshear, E., Bohlin, R., and Carlson, J. S. (2020, May). "Construction of Bounding Volume Hierarchies for Triangle Meshes with Mixed Face Sizes." In *2020 IEEE International Conference on Robotics and Automation (ICRA)*, pp. 9191-9195.

[37] Wiseman, Y., (2017), "Automatic Alert System for Worn Out Pipes in Autonomous Vehicles," *International Journal of Advanced Science and Technology*, Vol. 107, pp. 73-84.

[38] Duan, Z., Wang, N., Fu, J., Zhao, W., Duan, B., and Zhao, J., (2018). High precision edge detection algorithm for mechanical parts. *Measurement Science Review*, 18(2), 65-71.

[39] Wiseman, Y. and Grinberg I., (2016), "When an Inescapable Accident of Autonomous Vehicles is Looming," *International Journal of Control and Automation*, Vol. 9 No. 6, pp. 297-308.

[40] Wiseman, Y. and Grinberg I., (2016), "Autonomous Vehicles Should Not Collide Carelessly," *Advanced Science and Technology Letters*, Vol. 133, pp. 223-228.

[41] Wiseman, Y. and Grinberg I., (2016), "Circumspectly Crash of Autonomous Vehicles," *Proceedings of IEEE International*

Conference on Electro Information Technology (EIT 2016), Grand Forks, North Dakota, USA, pp. 382-386.

[42] Wiseman, Y., (2018) "Ancillary Ultrasonic Rangefinder for Autonomous Vehicles," *International Journal of Security and its Applications,* Vol. 12(5), pp. 49-58.

In: Advances of Machine Learning … ISBN: 978-1-68507-211-7
Editors: Pandian Vasant et al.

Chapter 10

CONTRIBUTION OF MACHINE LEARNING TO RAIL TRANSPORT SAFETY

Habib Hadj-Mabrouk*[*], PhD
Université Gustave Eiffel, Marne-la-Vallée, France

ABSTRACT

The seriousness and negative consequences of accidents and incidents in rail transport have forced all players in the rail system to set up an "experience feedback" system (REX) in order to improve operational safety. In this context, the knowledge of accidents and incidents results essentially from the contribution of lessons learned and experiences acquired. This process of analyzing and using experience data can be facilitated by the use of artificial intelligence techniques, especially machine learning, in order to understand the origins and circumstances of accidents and therefore propose solutions to avoid the reproduction of similar insecurity events. The approach proposed for the prevention of railway accidents is a Hybrid method built around three learning algorithms and one knowledge-based system (KBS):

- The acquisition of knowledge to gather knowledge of railway safety and in particular the scenarios of potential accidents,

[*] Corresponding Author's Email: habib.hadj-mabrouk@univ-eiffel.fr.

- Learning by classification of concepts to group accident scenarios into homogeneous classes such as the class relating to train collision or derailment problems.
- Rule-based machine learning (RBML) to automatically identify, from a base of historical scenarios (experience feedback), the relevant safety rules that are often difficult to extract manually from safety experts,
- Knowledge-based system (KBS). Production rules, previously induced by machine learning, are transferred to KBS to form the knowledge base of the safety assessment support tool.
- Case-based reasoning system (CBR). At the previous level, the KBS is used to evaluate safety at the highest level of the safety analysis hierarchy and can deduce a possible risk of accident not taken into account and likely to jeopardize the safety of the system and by therefore the safety of hardware and software equipment. This risk of accident requires the implementation of new prevention or protection measures during the various safety analyzes of hardware and software equipment (low level of the hierarchy). In this context, the CBR makes it possible to look for the most similar cases to this new risk of accident and proposes the appropriate measures.

Keywords: railway, artificial intelligence, classification, assessment, rule-based machine learning (RBML), case-based reasoning (CBR), Accident scenarios, functional safety analysis (FSA), security of critical software

INTRODUCTION

This article presents a contribution to the improvement of conventional methods of analysis and evaluation of railway safety. The objective is to use the experience of accident risk management to assist safety experts and possibly technical investigators to assess the safety level of a new rail transport system. This study relies heavily on machine learning methods and concerns two methods of rail safety analysis: functional safety analysis (FSA) and software error analysis (SEEA) methods. In spite of the undeniable interest of these methods, no method, alone, makes it possible to carry out an exhaustive safety analysis. The completeness of the safety analysis remains largely based on human intelligence and intuition. It is this fundamental point that has guided our study towards the use of artificial intelligence techniques to learn, from the experience of undesirable situations (failures), new knowledge likely to avoid at least the reproduction of historical accidents and incidents. The choice of an application-specific learning system is usually based on the identification of needs, the characteristics of available knowledge and the definition of the expected performance of the learning process. Thus, after having recalled the essential characteristics of the field and introduced our approach for the development of a tool to help in the evaluation of safety, we justify the choice of the learning systems selected with reference to all the properties required by the safety tool and the current offer perceived through the bibliography presented in the second paragraph. Referring to the bibliographic study, it can be seen that none of the learning systems considered satisfies to all the properties specified in the third paragraph.

We considered using the "Charade" (Ganascia, 1987) system for the generation of rules, the tool "ReCall" for case-based reasoning (CBR), but one is forced to develop a new system called "Clasca" (Hadj-Mabrouk, 2016) for the classification of accident scenarios. These three learning systems are complementary for the development of a tool to assist in the analysis and evaluation of railway safety. The results obtained to date are successively presented in the fourth and fifth paragraphs. The first classification module for accident scenarios is based on two main steps. An induction step for the construction of a description of each of the scenario classes and a deduction step for the search of the membership class of a new scenario. The second evaluation level, which is based on the joint use of the "Charade" system and an expert system, makes it possible to generate hazard recognition functions for each class produced by "Clasca." With the help of an expert system, these potential hazards are confronted with the hazards suggested by the manufacturer with a view to generating new dangers that could jeopardize the safety of the system. Faced with a particular danger constituting a new situation of insecurity, the CBR tool seeks to identify the solutions (or recommendations) necessary to guard against the danger previously generated. The sixth paragraph is devoted to the validation of the proposed methodology which took place in two phases. The first phase consisted of an overall assessment of the feasibility model based on its application to all the accident scenarios archived to date relating only to the risk of an accident: train collision. The second phase focused on an internal evaluation of the "Clasca" classification module based on a set of test scenarios provided by the experts. This validation revealed not only the interests but also the current limitations of the completed model and resulted in prospects for improvement presented in the seventh and final paragraph dedicated to the discussion of the results obtained.

Analytical Framework

Literature Review

In recent years and in the field of land and air transport, researchers and experts in the field have become increasingly interested in the application of artificial intelligence techniques to solve certain problems of aid the decision, such as the diagnosis of transport equipment, the management of maintenance operations, the analysis of driver behavior, the prediction of the deterioration of transport infrastructure, the planning and forecasting of traffic demand, control of traffic signals, control of air traffic, etc. For example, machine learning has been used for rail maintenance forecasting (Bergmeir et al., 2013), Expert Systems (fuzzy knowledge based) for rail traffic control (Fay, 2000), deep learning for the detection of lateral defects of the railroad (Santur et al., 2017) and neural networks for the detection of defects on the surface of rails (Faghih-Roohi et al., 2016). In railway transport applications, the Big Data Analytics (BDA) can be of a beneficial contribution in view of the large amounts of data generated by the transport system from sensors installed on the tracks, on the wagons or from the signaling equipment, monitoring and inspection equipment, communication systems, train monitoring systems, etc. A BDA can examine the collected data set in order to obtain useful information to explain for example the potential causes of degradation of the operation, the failure of certain track components and possibly safety equipment. As an example, we can mention the work on exploiting data relating to operation, maintenance and railway safety (Ghofrania et al., 2018), decision-making on rail maintenance (Thaduri et al., 2015), engineering and the management of railway applications (Attoh-Okine, 2014), the improvement of call reporting systems (Peter, 2018), the implementation of a predictive approach to the safety and maintenance of personnel (Hayward, 2018) and Siemens on the use of Big Data to build the Internet of trains (Marr, 2017). In the field of "data mining" and retrieval of relevant information from accident investigation reports, there are several techniques from various fields, such as information retrieval (IR), natural language processing (NLP), information extraction (IE), BDA approach and machine learning. The main goal is to explore plain text in order to extract relevant information for explanatory or decisional purposes.

In the field of railway safety, these methods are generally used to extract the presence of informative entities on the causes of accidents, recurrent accidents, to understand the causes of accidents, to find causal relationships from the investigation reports on accidents. To analyze reports of major railway accidents, (Williams et al., 2016) use the text mining techniques of probabilistic topic modeling and k-means clustering. The results of this study show that the types of recurring accidents are lane defects, wheel defects, level crossing accidents and switching accidents. Studies also show, through case examples (feedback), how the results of the textual search of stories can improve the understanding of contributing factors to rail accidents. Brown's paper (Brown, 2016) describes the use of text mining (with the combination of other techniques) to automatically discover the characteristics of railway accidents and to gain a better understanding of the factors contributing to these accidents. (LI et al., 2018) also applied the text mining method to the risk analysis for the safety of urban rail transport in China. The word frequency analysis and cluster analysis identified 15 safety risk factors from 156 accident reports. In the context of text mining, (Williams et al., 2018) use a comparison between Latent Semantic Analysis (LSA) and Latent Dirichlet Allocation (LDA) for railway accident text analysis. (Syeda et al., 2018) uses a Big Data Analytics (BDA) approach to analyze incident reporting to reduce safety risks in railway projects and operations. (Van-Gulijk, 2018) presents the case for IT transformation and Big Data for managing safety risks on UK railways. To investigate the causal link between causes and safety deficiencies in the rail industry, (Kanza et al., 2017) uses natural language processing (NLP) and machine learning. The objective is to reveal the presence of informative entities on the causes of rail accidents from the raw texts of accident investigation reports in the United Kingdom (published by the Railway Accident Investigation Directorate: RAIB). (Ghomi et al., 2016) applies data mining techniques based on association rules and classification algorithms to identify the severity factors of injuries caused by crossing accidents. Exploitation of the accident database shows that train speed, the age of vulnerable road users and sex are the most influential accident factors. In order to improve the identification of accidents, (Zhang et al, 2019) developed an approach based on machine learning, in order to distinguish secondary accidents. This text mining study shows that the classification model implemented is effective in identifying the keywords that characterize secondary collisions. (Heidarysafa et al, 2018) uses in-depth learning to analyze railway accident narratives to understand the causes of accidents and their corresponding descriptions in survey reports. The main goal is to help label accidents more accurately. (Gibert et al., 2017) uses deep learning to inspect the railroad. (Osama et al., 2019) proposes a machine learning model for near-accident prediction from observed vehicle kinematics data. (Chenariyan et al., 2019) presents recent applications of machine learning in railway maintenance.

Case-based reasoning (CBR) is attracting more and more attention from researchers and experts in the rail transport sector. CBR is a well-established field of research based on AI techniques and in particular machine learning. This mode of reasoning, which is based on the notion of similarity, focuses primarily on problem solving based on experience. It is a cognitive process of human reasoning that relies heavily on how people acquire a new skill based on their past habits and experiences. CBR means using and exploiting old experiences to understand, explain, interpret or solve new situations similar to similar past situations. The general principle of CBR is to treat a new problem (target case) by remembering similar past experiences (source cases). This type of reasoning rests on the assumption that if a past experience and new circumstances are sufficiently similar, then everything can be explained or applied to past experience (source cases) and remains valid when applied to the new situation which represents the new problem to solve. CBRs are increasingly used in industrial applications such as technical diagnostics, medical diagnostics, image processing, law, design, planning, and so on. In the field of transportation, our literature search covered three transport sectors: Air, road and rail. In the field of air transport we can cite, for example, the prediction of accidents and incidents (Zubair et al., 2012). In the road transport sector, the application of CBR is numerous: Transport planning (Khattak et al.,1996), management of traffic flows (Sadeka et al., 2001), control of urban intersections to avoid road congestion (Zhenlong et al., 2008), the analysis of road collisions (Li et al., 2005), the improvement of traffic in urban intersections by developing new signaling plans (Kofod-Petersen et al., 2014), the control of traffic flow at intersections (traffic control systems: TCS) (Louati et al., 2016), the diagnosis of the driver's stress level (Begum et al., 2012) or the modeling of the risk of driver fatigue (Zhong et al., 2017). Finally, in the rail transport sector, studies include the diagnosis of locomotive failures (Varma et al., 1999), the recovery of incident reports (Johnson, 2000), the prevention of rail operations incidents (Cui et al., 2005), the command of railway rescue (Emergency Relief Command) (Li et al., 2010), analysis of safety risks related to the operation of the metro (Lu et al., 2013), automatic train conduction to reduce travel time and save fuel consumption (De-Souza et al., 2016) and finally the diagnosis of failures of the rail switching system (Zhao et al., 2017).

All of this work clearly shows that AI, BDA, CBR and machine learning will likely have an increasing impact on the safety of rail transport.

Motivations, Objectives and Contributions with Respect the State of the Art

The process of analyzing rail safety can be broken down into several levels: System, automatisms, hardware and software. Each level of safety analysis has one or more safety methods (Figure 1). At the system level, the main method is the "Preliminary hazard analysis" (PHA) method (Hadj-Mabrouk, 2017a). At the level of automatisms, safety experts generally use the "Functional Safety Analysis" (FSA) method. The FSA aims to justify that the design architecture of the system is safe against potential accidents identified by the PHA. At a software level, it is a question of carrying out several methods related to Software Safety Analysis (SSA). The SSA is generally based on the "Software Errors and Effects Analysis" (SEEA) method as well as on critical code reads. At the hardware level, several safety methods relating to Hardware Safety Analysis need to be established (Hadj-Mabrouk, 2016): "Fault Tree Analysis" (FTA) and "Analysis of Failure Modes, their Effects and their Criticality" (AFMEC). The AFMEC method is usually completed by the "method of Combining Summarized Failures" (MCSF) also named Significant Failures Combination Search Method. The AFMEC method, which usually highlights simple failures, must be supplemented by studying combinations of failures that result in undesirable (or dangerous) events. Thus, the MCSF method, used in the extension of the AFMEC method, inductively determines such combinations of failures. Generally, it is noted for one or more modes of failure, that the effects on the system are identical. These failure modes are then grouped into fault sets called "Summary Faults" (SF). As part of our approach to analyzing and evaluating rail transport safety, we use this concept of Summary Failure (SF).

Our research is part of two complementary safeties analyzes: FSA and SEEA. The need to rationalize traditional approaches to safety analysis (FSA and SEEA), to improve the quality of accident risk analyzes and finally to help experts in problem solving and decision-making, has led us to the development of machine learning tools to suggest potential accidents and / or the most appropriate protective or preventive measures to protect against a particular risk. The railway safety analysis and assessment approach is organized around two main activities. The first activity is to extract, formalize and archive potential accident scenarios to develop a standard case library covering the entire safety problem. These dangerous situations are archived in two separate databases. The first base is the Functional Safety Analysis (FSA) is called "Historical Scenarios Knowledge Base" (HSKB). The second base is dedicated to software safety analysis (SEEA). The second activity aims at exploiting this stored historical knowledge in order to develop a safety analysis know-how that can help the experts to judge the comprehensiveness of the safety analyzes.

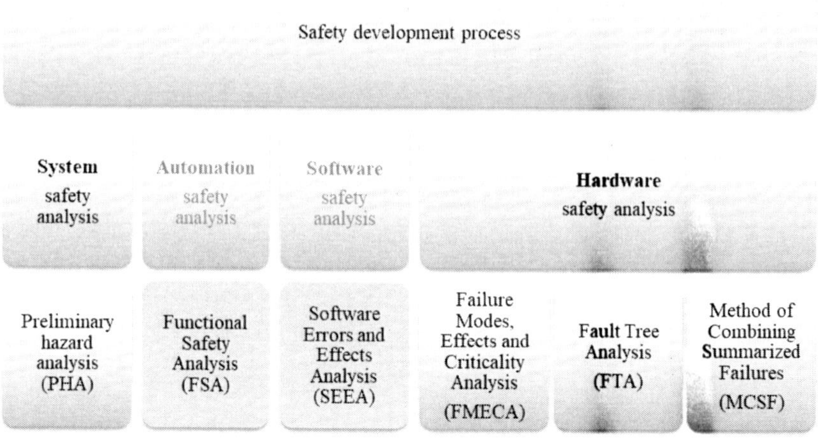

Figure 1. Main methods of analysis of rail transport safety.

This second activity is essentially based on the use of machine learning techniques. We use Rule-based machine learning (RBML) for evaluation of the FSA method and Case-based reasoning (CBR) for evaluation of the SEEA method. Despite the undeniable interest of artificial intelligence approaches presented in the previous paragraph, there is no comprehensive approach to meet all of our research objectives. The bibliographic study carried out on machine learning and in particular on CBR shows the absence of work on the use of CBR in the analysis and evaluation of the safety of critical software used in the rail transport sector. Faced with a complex such as safety and certification of transport systems, this approach successively calls upon the following methods: The acquisition of knowledge, learning by classification of concepts, Rule-based machine learning (RBML), Knowledge-based system (KBS) and Case-based reasoning system (CBR). It can be seen that none of the studied learning systems alone satisfies all these properties. However, if we break down our problem by distinguishing the classification activity from the evaluation activity of the accident scenarios, we can consider using the "Charade" (Ganascia, 1987) system for generation of production rules, but we are forced to develop a new classification system for accident scenarios: "Clasca" (Hadj-Mabrouk, 2016).

Method 1: Proposed Methodology for Evaluating the Functional Safety Analysis

As illustrated in Figure 2, the rail safety analysis and assessment methodology is organized in ten steps: Acquisition of safety knowledge, Pre-design, Learning, Classification, Validation, Study of convergence, Update of the HSKB database, Learning the Summarized Failures (SF), Deduction of SFs who are to be considered in the manufacturer's scenario and Validation by the safety Expert. The first seven steps are carried out by the scenario classification module "Clasca" and the last three steps concern the scenario evaluation module which is based on the "Charade" rule learning system. Very schematically, the first module of classification is concerned with carrying out a learning operation of the concepts and more precisely with descriptions characteristic of the historical scenarios resulting not only from the experience of the safety experts, but also from the safety files of the safety systems rail transport already certified and commissioned in France. The second evaluation module looks for regularities found in the scenario database and more precisely in each scenario class (identified by the previous module) in order to create a rule base for recognizing the presence of potential dangers involving the overall safety of the system. The first level of classification was presented in detail in (Hadj-Mabrouk, 2016; Hadj-Mabrouk, 2017b) and the second level of evaluation is presented in (Hadj-Mabrouk, 2018a, 2018b; Hadj-Mabrouk, 2019a). Therefore, we will present in the rest of this paper only the outline and the methodological indications allowing the reader to make the most of the articulation between these two learning modules. This new approach is presented below in three main phases: Acquisition of knowledge, classification and accident scenario assessment.

Acquisition and Modelling of Safety Knowledge

The knowledge acquisition phase has resulted in the development of a Historical Scenarios Knowledge Base (HSKB) that includes eighty scenarios of accidents or incidents related to collision risk such as the problem of redundancy switching, penetration on a busy canton, Improper Initialization, crossing a breakpoint in manual driving. All eighty scenarios were subsequently grouped by safety experts into several classes or family of scenarios such as the class "Initialization sequence" or the class "Emergency braking management." This HSKB, which forms the basis of the learning examples, will be exploited by the "Clasca" learning algorithm in order to find the membership class of a new scenario proposed by the transport system manufacturer. This HSKB database will also be exploited by the "Charade" learning algorithm to produce rules necessary for learning the Summary Fault (SF) recognition functions.

Classification of Accident Scenarios

The first level of analysis relates to finding the class to which a new scenario which has been suggested by the manufacturer belongs. "Clasca" is a learning system by researching classification procedures.

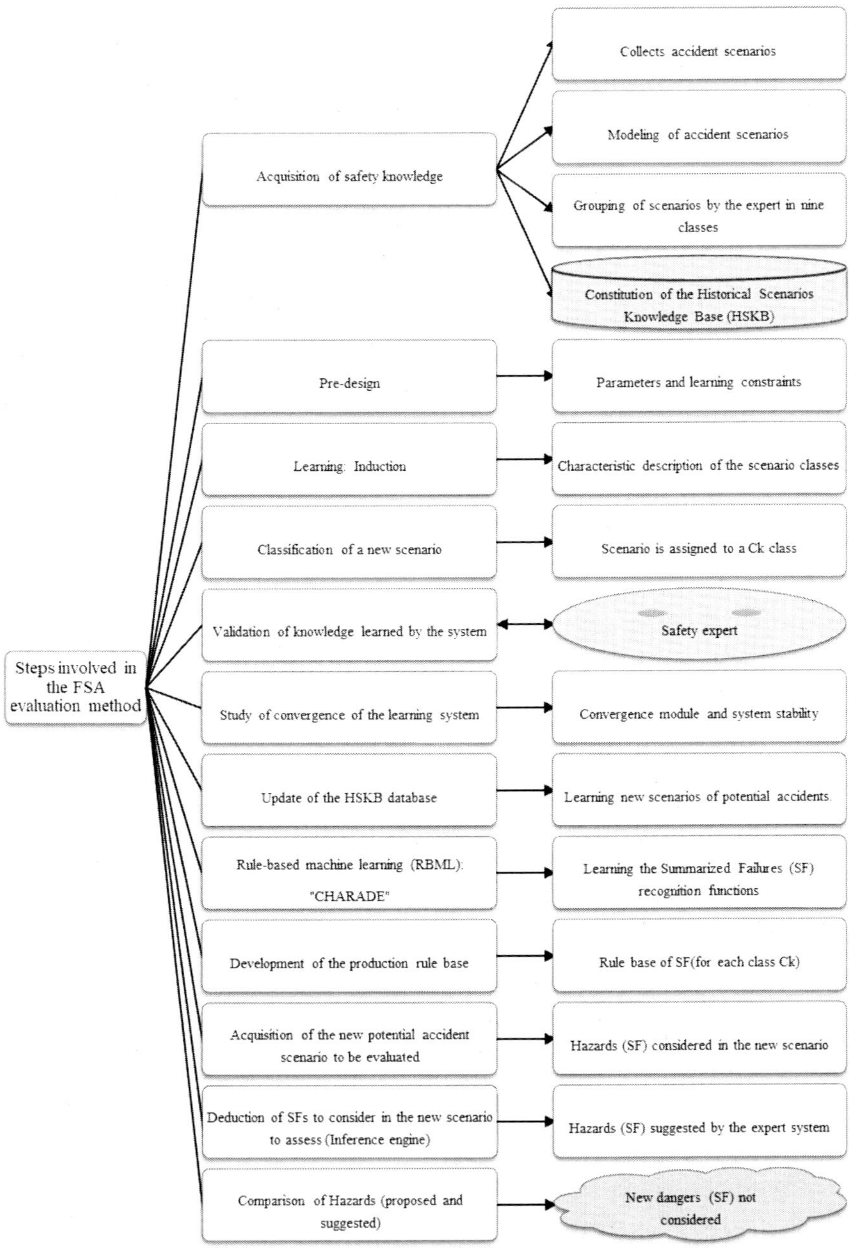

Figure 2. Steps involved in the process of assisting the evaluation of the SFA.

It incrementally develops conjunctive descriptions of historical scenario classes in order to characterize a set of insecurity situations and to identify a new scenario submitted for evaluation to the experts. The classification of a new scenario includes the following two major phases (Hadj-Mabrouk, 2016): 1) A characterization (or generalization) stage for constructing a description for each class of scenarios and 2) A deduction (or classification) stage to find the class to which a new scenario belongs by evaluating a similarity criterion. The descriptors of the new scenario are compared with the descriptions of the classes which were generated previously. In this stage a new example of a scenario is assigned to an existing class C_k. The classification phase requires the definition of a classification parameter called "adequacy rate" (T_{ad}) which measures the degree of resemblance between the new example E_i and each of the classes C_k of pre-existing scenarios.

Evaluation of Accident Scenarios

The second level of treatment considers the C_k class previously identified by the classification module in order to evaluate the consistency and completeness of the new accident scenario (scenario of the manufacturer that the expert seeks to examine). The method of analysis and evaluation of safety is centred on the summarized failures (SFs) which are involved in accident scenarios capitalized. The purpose is to automatically generate a recognition function for each SF associated with a scenario class. The SF recognition function is a production rule which establishes a link between a set of facts (parameters which describe a scenario) and the SF fact. In this context, it has proved to be inevitable to use a learning method which allows production rules to be generated from a set of historical examples (or scenarios). The specification of the properties required by the learning system and a review of the literature has led us to choose the "Charade" mechanism (Ganascia, 1987). Rule generation within "Charade" is based on looking for and discovering empirical regularities which are present in the entire learning sample (Figure 3).

If	Elements involved = mobile operator, Incident functions = instructions Elements-involved = operator in CC.
Then	summarized failures = SF11 (Invisible element on the zone of completely automatic driving), Elements involved = AD with redundancy, Hazard related functions =train localization, Geographical zones = terminus. [0]
If	Type of block = fixed block, Hazard related functions = initialization Incident functions = instructions
Then	summarized failures = SF10 (Erroneous re-establishment of safety frequency/high voltage,) Hazard related functions = Full control/High voltage permission Hazard related functions = alarm management, Hazard related functions = train localization. [0]

Figure 3. A sample of some rules generated by "Charade."

Thereby, a base of evaluation rules can be generated for each class of scenarios. The evaluation of a scenario involves two modules (Hadj-Mabrouk 2017b): 1) A mechanism for learning rules "Charade" which makes it possible to deduce SF recognition functions and thus generate a base of evaluation rules; 2) An inference engine which exploits the above base of rules in order to deduce which SFs are to be considered in the manufacturer's scenario.

METHOD 2: PROPOSED METHODOLOGY FOR ASSESSMENT THE SAFETY OF CRITICAL SOFTWARE

We have just presented the methodology implemented for help with FSAs. This paragraph is devoted to the presentation of a second complementary approach to help assess the safety of critical software and in particular the evaluation of SEEA. As shown in Figure 4, generally preliminary hazard analysis (PHD) can identify all potential accidents of the system such as collision, derailment of a train. As for the FSA, for each potential accident, it proposes to provide the functions and equipment needed to guard against these accidents. Our approach to safety assessment is organized around two closely related phases. To improve the FSA we proposed a new approach based in particular on the learning of the rules of production with a view to generate potential hazards (SF) not considered by the system designer. In the face of a particular danger, constituting a new situation of insecurity, this second approach to analyzing software safety focuses on identifying the solutions (or recommendations) necessary to cover the danger previously generated. This second evaluation phase, which is based on the use of case-based reasoning (CBR), aims to look for the most similar and analogous case in the historical accident and incident database (source case base) the problem to be addressed (target case) in order to propose the appropriate measures to avoid the occurrence of such a danger and consequently of a potential accident. Very schematically, in the context of the CBR, a case is considered a problem with his solution as well as procedures allowing a justification of the decisions made on the way the solution was generated.

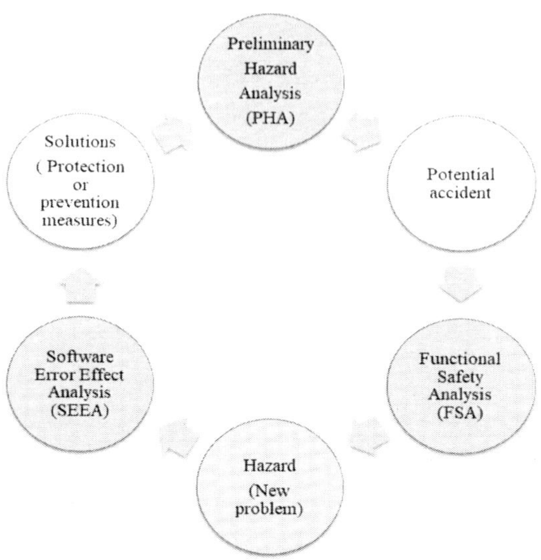

Figure 4. Safety analysis and assessment process.

In order to show the interest of machine learning and more precisely CBR in the field of the safety of railway transport, we have developed a tool to aid the assessment of the safety level of critical software are detailed in Figure 5 in nine steps presented in the following paragraphs. As shown in Figure 5, in front of each step of the proposed methodology, we presented the result obtained. For example, Step 1 on knowledge acquisition and modelling allowed the development of a generic SEEA representation model. This model is based on eight characteristic parameters: The investigated system, the investigated subsystem, the investigated module, the envisaged error (family, class, type), the safety criterion infringed by the error, the feared hazard, the type and severity of possible damage and finally the means of detecting the error and protecting against it. Using the above model we built up a library of 250 cases (examples). These historical examples of SEEAs were drawn from two guided transport systems: MAGGALY and the TVM 430 for the Nord TGV. The mock-up has been implemented using the "ReCall" software, marketed by "ISoft" firm, which generates CBR process. The mock-up has four main modules: A man/machine interface for inputting, updating and consulting knowledge relating to SEEA, a representation and acquisition module for SEEA sheets, a knowledge base containing 250 examples of SEEA (experience base), and a case-based reasoning process (implemented by the Recall software). The main components of this CBR process are a mechanism which indexes (or characterizes) target cases and a mechanism which finds similar cases (reference cases) and collects them together. The following paragraphs show, through an example the use of this mock-up, which requires going through the nine following stages (Hadj-Mabrouk, 2019b, 2020, 2021):

1. Acquisition and modelling of knowledge,
2. Definition of the description language of the SEEA examples,
3. Development the SEEA case base,
4. Parameterization and Calibrating of the CBR process,
5. Entering the new SEEA target case for evaluation,
6. Indexing of the SEEA case base,

7. Extraction of similar SEEA cases,
8. Adaptation of extracted cases (source cases),
9. Updating the SEEA base.

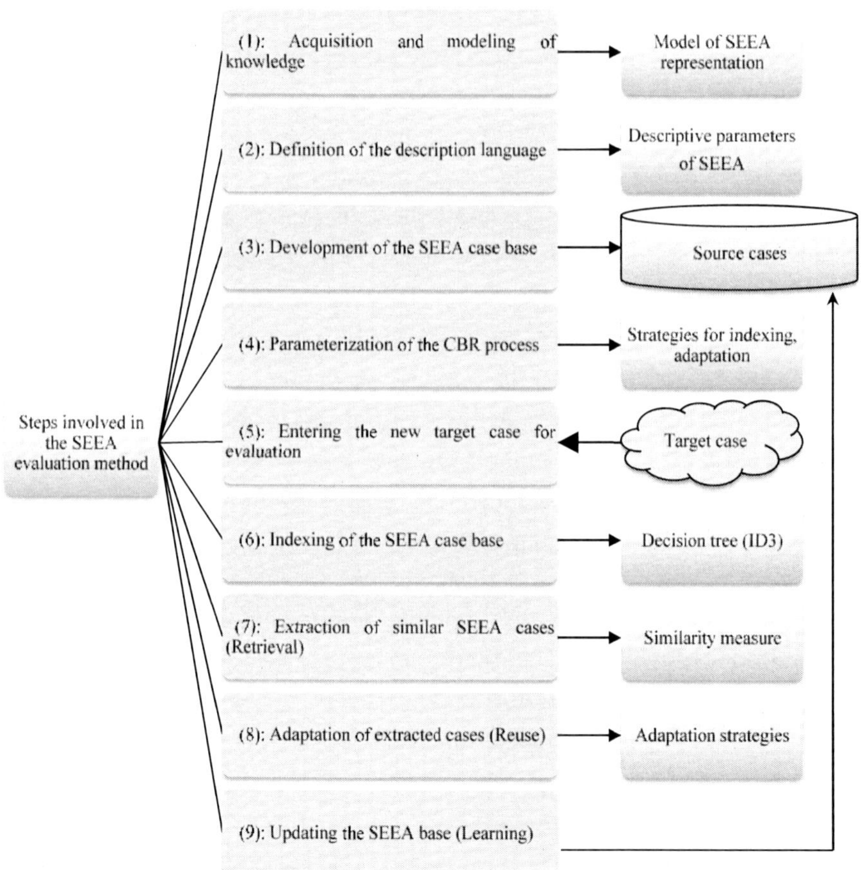

Figure 5. Approach for acquisition, modeling, capitalization and assessment of SEEA.

Acquisition and Modelling of Knowledge

The first step of the study is devoted to the research and identification of descriptors and characteristic parameters to represent and formalize the

SEEA. After a second step of data collection necessary to list the possible values taken by each parameter (or descriptor), the third step proposes, a formalism of representation of documents SEEA. Finally, on the base of this formalism, which constitutes the basic language of SEEA representation, the fourth stage of the study focuses on building the case base that currently comprises 224 cases, each of which represents a particular situation that is contrary to safety (Problem) and one or more preventive measures or corrective measures to guard against, avoid, reduce, or permanently eliminate the potential risk envisaged (Solution). To build this model and in order to show the feasibility of the study, we examined the SEEA relating only to two rail transport systems already certified and put into circulation in France: the automated system "MAGGALY" and the system TVM (track-to-train transmission) of the "LGV-Nord." At the end of this review, we finally proposed a first SEEA representation model that relies heavily on the manufacturers' practices and our experience in the field of railway safety. This formalism is based on eight characteristic parameters: Studied system, subsystem studied, module studied, error envisaged (family, class, type), safety criterion not respected by the error, dreaded event, type and gravity of the damage, barrier and means for detecting the error. This model proposes a methodological framework for preparing SEEA files and thus contributes to ensuring the quality of future analyzes.

Definition of the Description Language of the SEEA Examples

This step allows you to enter the description language of an SEEA based on the eight descriptors listed above. A descriptor is a couple (attribute, value). All attributes are symbolic. Three types of descriptors could be distinguished: Enumerated descriptors, multi-valued descriptors and unknown descriptors.

Developing the SEEA Case Base

It's about creating cases by assigning a value to each attribute of the description language. This case base may subsequently be modified or consulted. The acquisition of the target case is done by entering the value or values of the different attributes. During this case base construction step, the concept descriptor "dreaded event" is left unknown because it represents the solution we are looking for in the case base.

Parameterization of the CBR Process

During this step, the user must set different parameters to configure the CBR process. These choices concern both the descriptor that will represent the solution of the problem and the strategies of indexing, matching or adaptation. During this step, the user must set the following parameters:

- The descriptor "concept": The user must choose from all the descriptors the one that will represent the solution of the problem. In our example, the descriptor "concept" is the descriptor "dreaded event." The problem, meanwhile, will be characterized by all the other descriptors.
- Indexing strategies: In general, we use indexing rules that make it possible, on the one hand, to organize the case memory and, on the other hand, to express the relevant characteristics of the entries (the target cases) in terms of indexes. The process of extracting or choosing the source case strongly depends on the quality of the organization of the case memory. In our example, we construct the hierarchy by taking into account all the descriptors and by imposing the descriptors "studied system" and "studied subsystem," in this order, as first and second level of the decision tree. Then, the choice between the remaining descriptors for the

next levels will be done by a decision tree classification algorithm: Quinlan ID3 algorithm (Quinlan, 1986).

- Matching strategies to search for similar cases: Given a new problem to be solved (target case), it is, from a known case base (source cases), to find the most similar case (s) and relevant to solve the new problem. In this step, we generally use matching rules or similarity measures such as the "connective model" which imposes on each of the characteristics of the target case to be sufficiently close to all those of the source case or else "the disjunctive model" which evaluates the source case on its particularity closest to that of the target case. In this case, a source case will be considered acceptable if it is very close to the target case on at least one relevant characteristic, regardless of the value of the others. Most CBR systems evaluate the similarity of two cases by accounting for their common characteristics: This is the Euclidean distance. In our sample application, the user can intervene in several ways in calculating the similarity between two attributes. It can possibly specify the descriptors which will not have to be taken into account during the computation. It can also give a weight vector to indicate the relative importance of a descriptor over others. In our example, we chose to extract only the 10 most similar cases, and to give a weight equivalent to all the descriptors.
- Adaptation strategies for reuse of similar cases: Generally, there are two possibilities: 1) if the case found in the database (source case) is identical to the new problem to be solved (target case), then the solution of the problem is immediate; 2) Either the case found presents a certain similitude (or analogy) with the new case, then an adaptation procedure is necessary whose objective is to adapt the solution found to the need of the new situation (target case). Thus, in the first hypothesis, we apply directly the solution found and in the second hypothesis, we must find a suitable technique to adapt the recovered solution and include it in the new problem. In our sample application, to date, the tool does not offer

a real adaptation method, but allows the user to program his own methods with daemons. Currently, this adaptation can be done either implicitly by the safety domain expert, by comparing cases similar to the target case, or by the voting technique. In this second case, the value of the attribute to be adapted is calculated on all the similar cases by a vote weighted by the percentage of similarity of each case. For example, if a case C has 3 descriptors of which 2 are 100% similar to the target case and the third descriptor has no similarity (0%), then case C will be similar with the target case at 66%. If all the descriptors are of equal weight: $(100 \times \text{Descriptor weight } 1 + 100 \times \text{Descriptor weight } 2 + 0 \times \text{Descriptor weight } 3)/3 = 66$.

Entering the New SEEA Target Case for Evaluation

The acquisition of the target case is done by entering the value or values of the different attributes. W will leave the concept descriptor "dreaded event" unknown because it represents the solution we are looking for in the source case base.

Indexing of the SEEA Case Base

After developing the SEEA case representation mode, i.e., the description of the problem and the solution in the form of descriptors (attribute/value), it is then necessary to build a model for organizing and indexing the memory. The best way to build it is to use the decision tree method. Decision tree consists of nodes corresponding to the attributes of the selected objects and branches characterizing the alternative values of these attributes. The leaves of the tree represent the sets of objects of the same class of objects. The construction of decision trees is a top down generalization approach. The ID3 of QUINLAN algorithm (Quinlan, 1986) is a typical case of a downward approach. ID3 uses a heuristic search strategy, according to the gradient method, by optimizing a numerical criterion called gain of information which is based on the entropy of SHANNON developed in the early 1940s by Claude Shannon (Shannon, 1948).

Extraction of Similar SEEA Cases

The search for SEEA cases similar to the target case, is broken down into two filtering and selection stages that use static and dynamic indexes. There are different ways to determine the characteristics of indexes: All characteristics, some characteristics, the most discriminating characteristics, etc. In our application we adopted a similarity search based on the set of characteristics. In our application example, from the historical case base (source cases), it is a question of finding the SEEA cases most similar to the SEEA cases to be evaluated (target case) and who share the most important characteristics.

Adaptation of Extracted Cases (Source Cases)

In our application case, the "ReCall" tool used to demonstrate the feasibility of the proposed approach does not yet propose relevant adaptation strategies. To date, the adaptation phase is still assigned to the user and in particular to the safety expert. With the screen presented in Figure 6, the user can consult the value taken by the concept attribute "dreaded event" in each similar case and choose himself the value to give to the "concept" attribute for the target case. The user can also use the voting technique. In our example, the tool proposes a single value for the attribute "dreaded event": Train collision. Thus, the domain expert can adapt the most similar case (proposed by the tool) by assigning the "Feared Event" concept the value "Collision" as a solution to the problem.

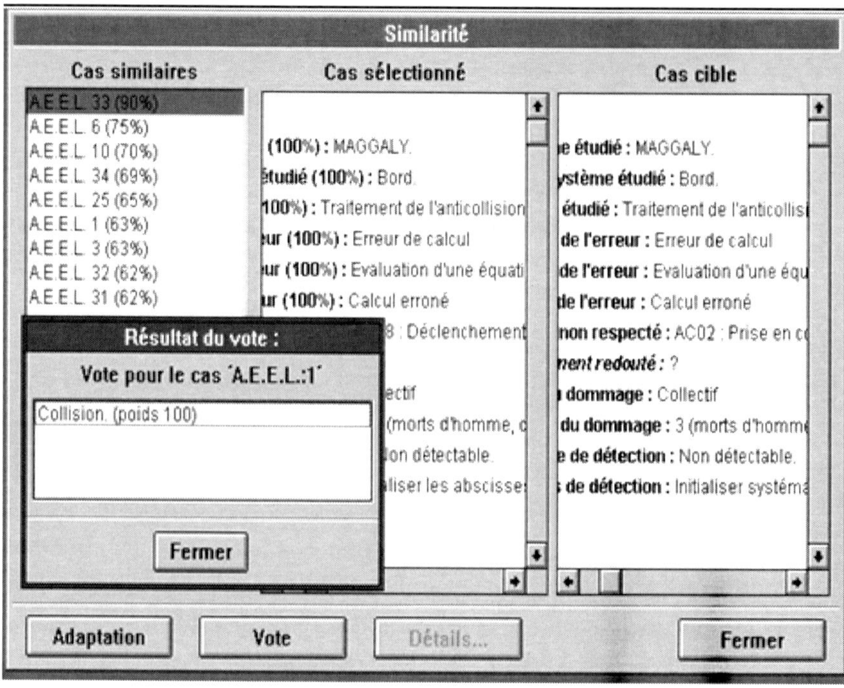

Figure 6. Example of the reference cases consultation and the vote technique use.

Updating the SEEA Base

This last step of updating knowledge is to perform the automatic learning by adding the appropriate target case in the SEEA historical case base. In the "ReCall" software, this learning is not incremental since the new case will be integrated into the hierarchy without it being reconstructed. It is up to the user to take the initiative to revive the indexing of the case base. Therefore, during this phase of the CBR cycle, it is wiser that the new case with its new solution is validated by the domain expert before being added to the case base (source cases).

RESULTS AND DISCUSSION

Validation

The validation of the proposed methodology took place in two phases. The first phase consisted of an overall assessment of the feasibility model based on its application to all the accident scenarios archived to date relating only to the risk of an accident: train collision. The second phase focused on an internal evaluation of the "Clasca" classification module based on a set of test scenarios provided by the experts. This validation revealed not only the interests but also the current limitations of the completed model and resulted in prospects for improvement. The quality of learning is measured by assessing the validity and usefulness of the knowledge learned. An acquaintance is valid if it is adequate and consistent with what is already known in the field of railway safety. It is useful if it helps to achieve the defined goals. The control of these two criteria is the responsibility of safety experts. To test the prediction capacity of the system and therefore the representativeness of the results obtained by the learning system, there are two solutions: either the expert has a list of control cases constituting the basis of test examples, or the all initial data available is shared in "learning base" and "test base" according to a distribution law given by the domain expert. Initially estimated at several hundred accident scenarios, the number of examples acquired so far is 80 scenarios for the learning base and 13 for the test (validation) scenarios. While this downward revision of the number of scenarios does not in any way affect the quality of the basis of examples of learning dedicated to collision risk, the problem of assessing the performance of the "Clasca" learning module arises. Table 1 presents the evaluation results obtained for the thirteen test scenarios. The classification results of these thirteen scenarios can be interpreted as follows:

- 9 scenarios are "well classified": they are classified in the same way by the system and the expert.

- 2 scenarios are "unclassified": their class matching rate is lower than the current similarity threshold and the system does not offer any class of membership. Since the similarity threshold is evaluative, these scenarios are momentarily put on hold and may be reclassified later. For one of these cases, the expert proposed a class and for the other case, he agreed with the decision of the system by not classifying the scenario.
- 2 scenarios are "misclassified": the classification of the system does not approve the expert. The main cause of this failure comes from the non-representative of the class descriptions learned on the basis of a small number of scenarios.

Although this first evaluation demonstrates the validity of the approach implemented on the "Clasca" model, a more detailed evaluation is needed to measure and eventually optimize the system's performance. For this, it is imperative to enrich the base of learning examples with additional scenarios relating to the risk of collision as well as scenarios relating to all other risks of accidents such as derailment. The main originality of "Clasca" lies in taking into account the non-monotone incrementality. This raises the problem of the influence of the order in which the examples are considered. Learning in CLASCA is non-monotonous: the description of a class learned at time t is called into question at time t + 1 and the frequency of appearance of a descriptor is sometimes increasing, sometimes decreasing. Although better suited to noisy data, non-monotonous learning does not guarantee the convergence of the process. In "Clasca," the convergence problem is controlled by the introduction of an evolutionary similarity threshold. The non-monotony of learning influences the sensitivity of "Clasca" to the order of arrival of the scenarios. An example that is integrated in a class at time t is not guaranteed to return to the same class at time t + n because of the evolution of the description of the class. This is even truer at the beginning of learning, when the description of the class is not yet well typed. To date, we have admitted to a first degree that the involvement of the expert in the control of classification results can circumvent this problem. A study on this point will however have to be conducted.

Table 1. Result of the validation of the module "Clasca"

Test scenarios	Prediction of "Clasca"		Decision Expert	Comparison of results
	Class	Adequacy rate		
SV1: Divergence between output control and input control	C7	1.00	C7	Well ranked
SV2: Premature registration of an item on a section	C7	0.77	C3	Misclassified
SV3: Late discrimination involving follower identification	C7	0.99	C7	Well ranked
SV4: Transmission fault between two adjacent PAs	C1	0.88	C1	Well ranked
SV5: Splitting a multiple unit on a DN	C7	1.00	C7	Well ranked
SV6: Three elements train	C7	0.83	C9	Misclassified
SV7: Lifting of the FU at zero speed after an evacuation request	?	< SS	C4	Unclassified
SV8: Manual driving on a section in evacuation request	C9	0.78	C9	Well ranked
SV9: Needle movement in manual control	?	< SS	?	Unclassified
SV10: Localization fault and needle movement	C3	0.94	C3	Well ranked
SV11: Reverse order of registration	C8	0.76	C8	Well ranked
SV12: Train in manual operation not respecting distance instructions	C9	0.91	C9	Well ranked
SV13: Counter-sense target	C6	0.83	C6	Well ranked

Discussion

In order to rationalize and reinforce conventional approaches to safety analysis and assessment, we have agreed to use machine learning techniques and in particular learning classification procedures (module "Clasca"), Rule-based machine learning (module "Charade"), case-based reasoning (module "ReCall") and knowledge-based system (KBS). The main objective consists, from a set of data in the form of accident scenarios or incidents experienced on rail transport systems (experience feedback), to exploit by automatic learning this mass of data (FSA and SEEA) in to stimulate the imagination of safety experts and assist them in their difficult task of analyzing and evaluating the safety. The knowledge acquisition phase ultimately culminated in the implementation of a conceptual FSA and SEEA representation model that provides a methodological framework for safety experts. Based on this model, we acquired 224 cases of SEEA and 80 accident scenarios involved in AFS. When it comes to machine learning, our work is part of supervised learning. Indeed, the presence of the safety expert is essential to ensure effective and relevant learning. To demonstrate the feasibility of the proposed approach, we used three learning systems. The first "Clasca" which is an inductive and incremental learning algorithm allows grouping and classifying the historical accident scenarios. The second "Charade," which was kindly provided by the Prf. Jean-Gabriel Ganasca (LIP6-Jussieu-Paris 6), strives to look for regularities present in a class of accident scenarios (proposed by the "Clasca" module) in order to produce a rule base that can be exploited by a expert system. For the third case-based reasoning algorithm (CBR), we used a CBR generator called "ReCall" from ISOFT. Despite the undeniable interest of these tools "Clasca," "Charade" and "ReCall," several shortcomings were noted during the evaluation phase. As stated above, for the "Clasca" system, it is necessary firstly to enrich the learning base so that it is representative of the field of railway risk management. With regard to the case-based reasoning system "ReCall," several shortcomings have been noted in particular for methods for calculating similarity, coping strategies and processing missing values (noisy data). Finally, the rule learning system "Charade," despite its undeniable interest, some rules generated are not of direct interest to assess safety. It is therefore essential to check, with the help of the domain expert, the veracity and the relevance of certain rules.

Contributions and Benefits of Results

The approaches proposed within the framework of this chapter make it possible to renew and strengthen the usual methods of certification and assessment of railway safety. They show the feasibility and interest of machine learning techniques and in particular of case-based reasoning to aid in the analysis of rail transport safety. The models and tools developed make it possible to help and assist not only safety experts during the certification phase of a new transport system, but also technical investigation bodies when researching accident scenarios or incidents likely to jeopardize the safety of travellers.

Limitations

Currently, project is at the mock-up stage. Initial validation has demonstrated the interest of the suggested approaches, but improvements and extensions are required before they could be used in an industrial environment or adapted to other areas where the problem of investigating safety arises. These improvements include the improvement of the adaptation strategies of the solutions proposed by the system, the enrichment of the basis of accident scenarios to cover the whole problem and finally, it is necessary to construct an integrated version of a prototype in order to finalize the results of the demonstration model.

Directions/Future Research Work

The currently developed classification learning systems produce decision trees, conceptual hierarchies or classification rules. Updating a concept hierarchy or decision tree is a very complex process that requires significant computing time. Obtaining relevant classification rules presupposes the existence of an almost exhaustive base of learning examples. In the context of our study, we do not have such a basis. This is why, within the framework of the "Clasca" system, we have opted for a classification which is based on the descriptions of the induced classes and on an adequacy criterion. The relevance of classification decisions is reinforced as the descriptions of the classes are enriched. When "Clasca" converges, class descriptions become stable and relevant, and classification decisions are reliable. Ultimately, the objective is therefore to improve "Clasca" by generating classification rules as soon as the class descriptions converge towards stable values. The description of a class is initially based on a numerical or statistical processing which distinguishes the characteristic descriptors from the "noise" descriptors. In the hypothesis of convergence of the system, only the characteristic descriptors of the classes are used to form symbolic classification rules. These rules can be used by an expert system for the classification of a new scenario. In this case, the premise of the rule contains the characteristic description (conjunctions of descriptors) and the conclusion of the rule contains the label of the class (docking of trains, redundancy switching, initialization sequence ...). To date, the proposed approach relating to the generation of classification rules is being implemented for evaluation.

CONCLUSION

Based on the examination of the manufacturer's safety file, the expertise of the experts, the history of the incidents that occurred and taking into account a certain number of constraints (safety objectives), the certifying expert relying on safety analysis tools and methods are required to develop a certification report. As long as the certifying expert demonstrates through accident scenarios that the safety objectives are not achieved, the designer must review and improve the transport system in terms of safety. This iterative process of safety evaluation, which relies on several modes of reasoning (inductive, deductive, by analogy) and uses several types of knowledge (incomplete, empirical, qualitative), confirm that a conventional computing solution is inappropriate and that the use of Artificial Intelligence (AI) techniques seems more appropriate. For our work, we used three specific aspects of the field of AI: knowledge acquisition, machine learning and Knowledge Base Systems (BSC). Despite all the care taken in the knowledge acquisition phase and the large number of knowledge extraction sessions (about thirty sessions) as well as the use of several knowledge gathering techniques (interviews, questionnaires, protocol analysis, etc.), the knowledge acquisition phase only led to a precise characterization of the accident scenarios (set of descriptive parameters) before encountering the difficulty of revealing the strategies, heuristics and reasoning mechanisms of the experts . This obstacle to the acquisition of the expert approach is due in particular to the intuitive and evolving nature of the knowledge. The acquisition of knowledge has made it possible to precisely characterize the area of expertise, to identify all of the fundamental concepts and to reframe the study's objectives to take account of realities. It led to the development of a "generic" model for the representation of accident scenarios. The scenarios collected to date concern the risk of collision. As a result, the use of other means of automatic knowledge acquisition (machine learning) has become necessary.

In order to assist the experts in the evaluation of safety studies and in particular the FSA and SEEA, the choice of one or more learning techniques is required. Referring to the bibliographic study, it can be seen that none of the learning systems considered alone satisfies all these properties. However, if we break down our problem by distinguishing the classification activity from the evaluation activity of the accident scenarios, we considered using the "Charade" system for the generation of rules, the tool "ReCall" for case-based reasoning (CBR), but one is forced to develop a new system called "Clasca" for the classification of accident scenarios. The first classification module for accident scenarios is based on two main steps: 1) a characterization step (or generalization) for the construction of an intentional description of each of the scenario classes. This step operates by detecting similarities in a set of historical scenarios pre-classified by the domain expert; 2) a deduction step (or classification) for the search of the class of membership of a new scenario by evaluation of a criterion of adequacy. The second evaluation level, which is based on the joint use of the "Charade" system and an expert system, makes it possible to generate hazard recognition functions for each class produced by "Clasca." With the help of an expert system, these potential hazards are confronted with the hazards suggested by the manufacturer with a view to generating new dangers that could jeopardize the safety of the system. Faced with a particular danger constituting a new situation of insecurity, the CBR tool seeks to identify the solutions (or recommendations) necessary to guard against the danger previously generated.

REFERENCES

Attoh-Okine, N. (2014). Big data challenges in railway engineering. *IEEE International Conference on Big Data (Big Data)*, 27–30 Oct. 2014, Washington, DC, USA.

Begum, S. Ahmed, MU. Funk, P. et al. (2012). Mental state monitoring system for the professional drivers based on Heart Rate Variability analysis and Case-Based Reasoning. *Federated Conference on Computer Science and Information Systems (FedCSIS)*, 9–12 Sept. 2012, Wroclaw, Poland.

Bergmeir, C. Sáinz, G. Bertrand, CM. et al. (2013). A Study on the Use of Machine Learning Methods for Incidence Prediction in High-Speed Train Tracks, In: Ali M, Bosse T, Hindriks KV, Hoogendoorn M, Jonker CM, Treur J (eds.), Recent Trends in Applied Artificial Intelligence, *IEA/AIE 2013, Lecture Notes in Computer Science*, Springer, Berlin, Heidelberg, 7906, 674–683.

Brown, D-E (2016) Text Mining the Contributors to Rail Accidents, in *IEEE Transactions on Intelligent Transportation Systems*, Feb. 2016, 17(2), 346-355.

Chenariyan Nakhaee, M. Hiemstra, D. Stoelinga, M. and van Noort, M. (2019). The Recent Applications of Machine Learning in Rail Track Maintenance: A Survey. In: Collart-Dutilleul S., Lecomte T., Romanovsky A. (eds) Reliability, Safety, and Security of Railway Systems. Modelling, Analysis, Verification, and Certification. RSSRail 2019. *Lecture Notes in Computer Science*, vol 11495. Springer.

Cui, Y. Tang, Z. and Dai, H. (2005). Case-based reasoning and rule-based reasoning for railway incidents prevention. *Proceedings of ICSSSM*

'05. 2005 International Conference on Services Systems and Services Management, June 2005, Chongquing, China, pp. 13–15.

De Souza, VDM. Borges, AP. Sato, DMV. et al. (2016). Automatic knowledge learning using Case-Based Reasoning: A case study approach to automatic train conduction. *International Joint Conference on Neural Networks (IJCNN)*, July 2016, pp. 24–29.

Faghih-Roohi, S. Hajizadeh, S. Núñez, A. et al. (2016). Deep convolutional neural networks for detection of rail surface defects. *International Joint Conference on Neural Networks (IJCNN),* July 2016, Canada, pp. 24–29.

Fay, A. (2000) A fuzzy knowledge-based system for railway traffic control. *Eng Appl Artif Intel* 13(6), 719–729.

Ganascia, J-G. (1987). *Agape et Charade: deux mécanismes d'apprentissage symbolique appliqués à la construction de bases de connaissances* [*Agape and Charade: two symbolic learning mechanisms applied to the construction of knowledge bases*]. Thèse d'État, Université Paris-sud, France 1987.

Ghofrania, F. He Q. Goverde, R. et al. (2018). Recent applications of big data analytics in railway transportation systems: A survey. *Transport Res C-Emer* 90, 226–246.

Ghomi, H. Bagheri, M. Fu L. and Miranda-Moreno, L-F. (2016). Analyzing injury severity factors at highway railway grade crossing accidents involving vulnerable road users: A comparative study, *Traffic Injury Prevention*, 17(8), 833-841.

Gibert, X. Patel, VM. and Chellappa, R. (2017). Deep multitask learning for railway track inspection. *IEEE Trans. Intell. Transp. Syst.* 18, pp. 153–164.

Hadj-Mabrouk, H. (2016a). CLASCA: Learning System for Classification and Capitalization of Accident Scenarios of Railway. *Journal of Engineering Research and Application*, 6 (8), 91-98.

Hadj-Mabrouk, H. (2016b). Machine learning from experience feedback on accidents in transport. *Published in 7th International Conference on Sciences of Electronics, Technologies of Information and*

Telecommunications, doi: 10.1109/SETIT.2016.7939874, pp. 246-251.

Hadj-Mabrouk, H. (2017a). Preliminary Hazard Analysis (PHA): New hybrid approach to railway risk analysis. *International Refereed Journal of Engineering and Science*, 6 (2), 51-58.

Hadj-Mabrouk, H. (2017b). Contribution of learning Charade system of rules for the prevention of rail accidents. *Intelligent Decision Technologies,* 2017, 11(4), 477–485.

Hadj-Mabrouk, H. (2018a). A Hybrid Approach for the Prevention of Railway Accidents Based on Artificial Intelligence, In: Vasant P, Zelinka I, Weber GW (eds.), *International Conference on Intelligent Computing & Optimization*, pp. 383–394.

Hadj-Mabrouk, H. (2018b). New approach of assessing human errors in railways. *Transactions of the VSB - Technical University of Ostrava, Safety Engineering Series*, 13(2), 1-17.

Hadj-Mabrouk, H. (2019a). Contribution of artificial intelligence to risk assessment of railway accidents. *Journal of Urban Rail Transit*, 5(2), 104–122.

Hadj-Mabrouk, H. (2019b). Contribution of artificial intelligence and machine learning to the assessment of the safety of critical software used in railway transport. *Journal AIMS Electronics and Electrical Engineering*, 2019, 3(1), 33-70.

Hadj-Mabrouk, H. (2020). Application of Case-Based Reasoning to the safety assessment of critical software used in rail transport. *Journal Safety Science*, Vol. 131, November 2020, (104928).

Hadj-Mabrouk, H. (2021). Case-based reasoning for safety assessment of critical software. *Intelligent Decision Technologies*, 2020, 14 (4), 463–479.

Hayward, V (2018). *Big data & the Digital Railway*. Available from: https://on-trac.co.uk/big-data-digital-railway/.

Heidarysafa, M. Kowsari, K. Barnes, L-E. Brown, D-E. (2018). Analysis of Railway Accidents' Narratives Using Deep Learning. *International Conference on Machine Learning and Applications (IEEE ICMLA),* arXiv: 1810.07382 [cs.CL]. doi: 10.1109/ICMLA.2018.00235.

Johnson, C. (2000). Using case-based reasoning to support the indexing and retrieval of incident reports. *Proceeding of European Safety and Reliability Conference (ESREL 2000)*: Foresight and Precaution, Balkema, Rotterdam, the Netherlands, pp. 1387–1394.

Kanza Noor and Shirazi, Noor Ul Hassan and Naqvi, Asad Ali and Parkinson, Howard J. and Bamford, G. (2017). Big Data and Natural Language Processing for Analysing Railway Safety. In: *Innovative Applications of Big Data in the Railway Industry.* IGI Global Publishing, pp. 240-267.

Khattak, A. and Kanafani, A. (1996). Case-based reasoning: A planning tool for intelligent transportation systems. *Transport Res C-Emer* 4, 267–288.

Kofod-Petersen, A. Andersen, OJ. and Aamodt, A. (2014). Case-Based Reasoning for Improving Traffic Flow in Urban Intersections, In: Lamontagne L, Plaza E (eds.), Case-Based Reasoning Research and Development, *ICCBR 2014, Lecture Notes in Computer Science*, Springer, Cham, 8765, 215–229.

Li, J, Wang, J. Xu, N. Hu, Y. and Cui, C. (2018). Importance Degree Research of Safety Risk Management Processes of Urban Rail Transit Based on Text Mining Method, *Journal Information*, 9(2), 26.

Li, K. and Waters, NM. (2005). Transportation Networks, Case-Based Reasoning and Traffic Collision Analysis: A Methodology for the 21st Century, In: Reggiani A, Schintler LA (eds.), *Methods and Models in Transport and Telecommunications, Advances in Spatial Science*. Springer, Berlin, Heidelberg, pp. 63–92.

Li, X. and Yu, K. (2010). The research of intelligent Decision Support system based on Case-based Reasoning in the Railway Rescue Command System. *International Conference on Intelligent Control and Information Processing*, 13–15 Aug. 2010, Dalian, China.

Louati, A. Elkosantini, S. Darmoul, S. et al. (2016). A case-based reasoning system to control traffic at signalized intersections. *IFAC-Papers on Line* 49, 149–154.

Lu, Y. Li, Q. and Xiao, W. (2013). Case-based reasoning for automated safety risk analysis on subway operation: Case representation and retrieval. *Safety Sci* 57, 75–81.

Marr, B. (2017). *How Siemens Is Using Big Data And IoT To Build The Internet of Trains*. Available from: https://www.forbes.com/sites/bernardmarr/2017/05/30/how-siemens-is-using-big-data-and-iot-to-build-the-internet-of-trains/#2b7a4b6e72b8.

Osman, A. Hajij, M. Bakhit, P-R. and Ishak, S. (2019). Prediction of Near-Crashes from Observed Vehicle Kinematics Using Machine Learning. *Transportation Research Record*, doi:10.1177/0361198119862629.

Peter, H. (2018). *Making the railway safer with big data*. Available from: http://www.railtechnologymagazine.com/Comment/making-the-railway-safer-with-big-data.

Quinlan, JR. (1986). Induction of Decision Trees. *Mach Learn* 1, 81–106.

Sadeka, A. Smith, B. and Demetsky, M. (2001). A prototype case-based reasoning system for real-time freeway traffic routing. *Transport Res C-Emer* 9, 353–380.

Santur, Y. Karaköse, M. and Akin, E. (2017). A new rail inspection method based on deep learning using laser cameras. *International Artificial Intelligence and Data Processing Symposium (IDAP)*, 16–17 Sept. 2017.

Shannon, CE. (1948). A mathematical theory of communication. *Bell Syst Tech* J 27: 379–423.

Syeda, K. Shirazi, S. N. Naqvi, S. A. Parkinson, H. J. and Bamford G. (2018). Big Data and Natural Language Processing for Analysing Railway Safety: Analysis of Railway Incident Reports. In S. Kohli, A. Kumar, J. Easton, & C. Roberts (Eds.), *Innovative Applications of Big Data in the Railway Industry*. Hershey, PA: IGI Global. doi:10.4018/978-1-5225-3176-0.ch011, pp. 240-267.

Thaduri, A. Galar, D. and Kumar, U. (2015). Railway assets: A potential domain for big data analytics. *Procedia Comput Sci* 53, 457–467.

Van-Gulijk, C. Hughes, P. Figueres-Esteban, M. El-Rashidy, R. and Bearfield, G. (2018). The case for IT transformation and big data for safety risk management on the GB railways. *Proceedings of the*

Institution of Mechanical Engineers, Part O: Journal of Risk and Reliability, 232(2), 151–163.

Varma, A. and Roddy, N. (1999). ICARUS: Design and deployment of a case-based reasoning system for locomotive diagnostics. *Eng Appl Artif Intel* 12, 681–690.

Williams, T. Betak, J. and Findley, B. (2016). Text Mining Analysis of Railroad Accident Investigation Reports. *Proceedings of the 2016 Joint Rail Conference*. Columbia, South Carolina, USA. April 12–15, 2016. V001T06A009. ASME. https://doi.org/10.1115/JRC2016-5757.

Williams, T. and Betakbc, J. (2018). A Comparison of LSA and LDA for the Analysis of Railroad Accident Text, *Procedia Computer Science*, 130, 98-102.

Zhang, X. Green, E. Chen, M. Reginald, R. and Souleyrette, (2019). Identifying secondary crashes using text mining techniques, *Journal of Transportation Safety & Security*, doi:10.1080/19439962.2019.1597795.

Zhao, H. Chen, H. Dong, W. et al. (2017). Fault diagnosis of rail turnout system based on case-based reasoning with compound distance methods. *29th Chinese Control and Decision Conference (CCDC)*, 28–30 May 2017.

Zhenlong, L. and Xiaohua, Z. (2008). A case-based reasoning approach to urban intersection control. *7th World Congress on Intelligent Control and Automation*, 25–27 June 2008, Chongqing, China.

Zhong, Q. and Zhang, G. (2017). A Case-Based Approach for Modelling the Risk of Driver Fatigue, In: Shi Z, Goertzel B, Feng J (eds.), Intelligence Science I. ICIS 2017. *IFIP Advances in Information and Communication Technology*, Springer, 510, 45–56.

Zubair, M. Khan, MJ. and Awais, M. (2012). Prediction and analysis of air incidents and accidents using case-based reasoning. *Third Global Congress on Intelligent Systems*, 6–8 Nov. 2012, Wuhan, China.

In: Advances of Machine Learning … ISBN: 978-1-68507-211-7
Editors: Pandian Vasant et al.

Chapter 11

THE POWER OF VARIABLE FREEING AND VARIABLE SUM BOUNDS IN SOLVING THE LINEAR KNAPSACK PROBLEM

Elias Munapo*
Department of Statistics and Operations Research,
School of Economics Sciences, North West University,
Mafikeng, South Africa

ABSTRACT

This paper presents a technique of using variable freeing and variable sum bounds in solving the general linear knapsack problem. In the paper, the lower and upper bounds of a sum of variables for any linear knapsack model is accurately determined. The basic variable in the continuous optimal solution of the knapsack problem, which is also known as the variable with the largest weight, is freed of integral restrictions before being solved by the standard branch and bound algorithm. A linear knapsack problem with variable sum bounds and a freed basic variable is easier to solve by the standard branch and bound algorithm than its original form. There are so many important

* Corresponding Author's Email: emunapo@gmail.com.

applications of the knapsack problem and one of them is its application in the transportation sector.

Keywords: linear knapsack problem, variable sum bounds, freed variable and branch and bound algorithm

1. INTRODUCTION

Integer models have practical applications such as in transportation, capital budgeting, traveling salesman, facility location and scheduling. The linear knapsack model is a special case of linear integer models whereby there is only one constraint. The general linear integer model is NP Complete and is very difficult to solve. The general linear knapsack problem has only one constraint but it is still NP Complete and also very difficult to solve.

There are several types of knapsack problems which include zero-one, multiple-choice, collapsing zero-one, quadratic, set union, multi-dimensional, bounded, unbounded and multi-objective knapsack models. In the zero-one class of knapsack problems, variables assume 0s and 1s only and is the easiest class as compared to all other classes of knapsack models. The zero-one class comes from the fact that one can select an item or not. In this case, it is not possible to select a fraction of an item such as a human being, car or a day to do business. Please see [1] or [2] for more on zero-one knapsack problems.

The multiple knapsack problem is formed by combining single knapsacks into a group of knapsacks having different capacities. For this knapsack model, the objective is to assign each item to, at most, one of the knapsacks such that all capacity constraints are satisfied and that the total profit of all the items put into knapsacks is made maximum. You may see [3] or [4] or [5] for more on these multiple knapsack problems.

In the bounded knapsack problem there is a knapsack capacity and a set of items, where each item has a positive integer value, a positive integer weight, and a positive integer limit or bound on its availability. In this case, the objective is to select the number of each item type to add to the knapsack in such a way that the total weight is not violated and that the total value is a maximum [6].

In a quadratic knapsack model, the objective function is in the form of a quadratic function and is subjected to a set of linear constraints. The variables in this knapsack problem can either be zero-one or general integers. This type of knapsack problem which is in nonlinear form is very difficult to solve. Please see [7] or [8] or [9] or [10] for more on quadratic knapsack problems.

As for the set-union knapsack problem, it is a special case of the zero-one knapsack problem in which each item is a set of elements, whereby each item has a nonnegative value, and each element has a nonnegative weight. The weight of each item is determined by adding the weight of all the elements in the union of the items' sets. More on set-unions can be found in [11].

In the multi-objective knapsack model, the problem has many objectives in the same problem. In an organization for example, the objectives may include minimization of labour and transportation costs and then maximization of profits and these objectives must be achieved simultaneously [12, 13]. As for the multidimensional knapsack problem, profit is maximized while observing some capacity restrictions. See for example, [14-18] for more on multidimensional knapsack models.

2. Applications of the Knapsack Model

The knapsack model has practical applications in formulation and solution methods for tour conducting and optimisation of content delivery networks [19], home energy management [20], electrical power allocation management [21], cognitive radio networks [22], mining operation use [23], production planning [24], in selection of renovation actions [25], waste management [26] and relay selection in secure cooperative wireless communications [27].

3. THE LINEAR INTEGER PROBLEM

3.1. The General Linear Integer Problem

The general linear integer problem can be defined as given in (1)

Maximise $Z = c_1x_1 + c_2x_2 + ... + c_nx_n$,

Such that

$$
\begin{aligned}
&a_{11}x_1 + a_{12}x_2 + ... + a_{1n}x_n \le b_1,\\
&a_{21}x_1 + a_{22}x_2 + ... + a_{2n}x_n \le b_2,\\
&...\\
&a_{m1}x_1 + a_{m2}x_2 + ... + a_{mn}x_n \le b_m.
\end{aligned}
\tag{1}
$$

where a_{ij}, b_i and c_j are constants and x_j is integer, $i = 1,2,...,m$ and $j = 1,2,...,n.$

3.1.1. Continuous Optimal Solution of the General Integer Problem

Table 1 gives the continuous optimal solution. The variables in Table 1 can be arranged in so many ways.

We used this way for our convenience.

Since this is a continuous optimal tableau, then quantities in (2) and (3) are nonnegative.

$$\omega_1, \omega_2, \omega_3, ..., \omega_m \ge 0. \tag{2}$$

$$\beta_1, \beta_2, \beta_3, ..., \beta_m \ge 0. \tag{3}$$

Table 1. Continuous optimal tableau

Basic variables (x)					Nonbasic variables (s)					rhs
0	0	0	...	0	ω_1	ω_2	ω_3	...	ω_m	γ
1	0	0	...	0	α_{11}	α_{12}	α_{13}	...	α_{1m}	β_1
0	1	0	...	0	α_{21}	α_{22}	α_{23}	...	α_{2m}	β_2
0	0	1	...	0	α_{31}	α_{32}	α_{33}	...	α_{3m}	β_2
...										
0	0	0	...	1	α_{m1}	α_{m2}	α_{m3}	...	α_{mn}	β_m

In the continuous optimal tableau, γ is the objective value and α_{ij} is a constant. Both γ and α_{ij} can either be negative or positive. This problem is NP Complete and is very difficult to solve. We are not aware of any consistent and efficient approach for this problem.

The general approach for this problem is to rely on the strategy of first finding a continuous optimal solution and then use other approaches to reach the optimal integer solution. The common approaches are:

(1) cutting plane [28],
(2) branch and bound [29],
(3) branch and cut [30-32],
(4) branch and price [32, 34] and
(5) branch cut and price [35-37].

Cutting plane method: A cutting plane method is a method that iteratively reduces the feasible region of a linear program by the use of cuts until the desired corner point is integral.

Branch and bound method: A branch-and-bound method is a systematic enumeration of the solution space. The searching process results in a search tree and a branch and bound method explores the search tree until an optimal solution is found.

Branch and cut method: Branch and cut is a combination of the branch and bound and the cutting plane methods. In other words the branch and bound is run in such a way that cuts are added in the branches. Cutting planes are added to tighten the linear programming (LP) relaxations.

Branch and price method: This method is a hybrid of branch and bound and column generation methods. From the fact that for large problems, most columns will be non-basic and have their corresponding variables equal to zero in the optimal solution, then a majority of the columns are irrelevant for solving the problem at each node of the search tree and columns may be selected and added to the linear programming relaxation (LP relaxation).

Branch cut and price method: This is a method whereby cutting planes are used to tighten LP relaxations within a branch and price algorithm.

3.2. Knapsack Problem and Statement of Problem

The linear knapsack problem is a special case of the general integer problem.

Even though in this integer problem, constraints have been reduced to only one, it is still NP complete and also very difficult to solve.

Minimise $Z = c_1x_1 + c_2x_2 + ... + c_nx_n$.

$$\text{Such that: } a_1x_1 + a_2x_2 + ... + a_nx_n \geq b \qquad (4)$$

where x_j is integer, c_j and a_j are nonnegative constants.

3.2.1. Continuous Optimal Solution

Since there is only one constraint, the continuous optimal solution is made up of a single basic variable. Let this single basic variable be given by (5).

$$x_j^0 = I + f. \tag{5}$$

where x_j^0, is the basic variable, I is the integral part and f is the fractional part of the basic variable.

3.2.2. The Upper Bound of the Objective Value Z_0^B

The upper bound of the objective value (Z_0^B) is given as (6).

$$Z_0^B = c_j(I+1). \tag{6}$$

Then the objective function is related to this value as given in (7).

$$c_1x_1 + c_2x_2 + \ldots + c_nx_n \leq Z_0^B. \tag{7}$$

This bound is easy to determine. This is not the only way that can be used to obtain a bound.

An alternative way is to calculate the weight of each variable and then use the smallest weight as the upper bound.

From the constraint in (4), we use another variable y_j such that, $a_1y_1 + a_2y_2 + \ldots + a_ny_n \geq b$ and that each of these y variable is related to b by the way given in (8).

$$a_jy_j \geq b, \tag{8}$$

$$y_j \geq \frac{b}{a},$$

$$y_j = x_j^0.$$

where $y_j = x_j^0$ is the smallest integer greater than $\left(\frac{b}{a}\right)$ and $x_j \leq x_j^0$.

The objective bound Z_0^B can be approximated as given in (9).

$$Z_0^B = \text{Min}[c_1 x_1^0, c_2 x_2^0, ..., c_n x_n^0]. \tag{9}$$

Assumption used: When calculating the bound of each variable, it is assumed that all other variables are zero.

3.2.3. Variable Sum Bounds

The variable sum bounds $(\ell_1 \,\&\, \ell_2)$ can now be determined once the objective bound is known. These two bounds are calculated in such a way that they satisfy (11). The use of a variable sum bound is not a new idea, it was used in [38].

$$\ell_1 \leq x_1 + x_2 + ... + x_n \leq \ell_t. \tag{10}$$

Adjusting these limits to the integer will make them tighter. The two variable sum bounds may be determined by solving two linear programming models given in (11) and (12) separately.

Maximise $\ell_t = x_1 + x_2 + ... + x_n$,

Such that:

$$\begin{aligned} &a_1 x_1 + a_2 x_2 + ... + a_n x_n \geq b, \\ &c_1 x_1 + c_2 x_2 + ... + c_n x_n \leq Z_0^B, \end{aligned} \tag{11}$$

where $x_j \geq 0$.

Minimise $\ell_1 = x_1 + x_2 + \ldots + x_n$,

Such that:

$$a_1x_1 + a_2x_2 + \ldots + a_nx_n \geq b,$$
$$c_1x_1 + c_2x_2 + \ldots + c_nx_n \leq Z_0^B, \tag{12}$$

where $x_j \geq 0$.

The separate LPs (12) and (13) have the advantage that a multi-processor can solve them easily at the same time. It is also possible to combine these separate LPs into one LP as given in (13).

Maximise $\ell_t - \ell_1$,

Such that:

$$a_1x_1 + a_2x_2 + \ldots + a_nx_n \geq b,$$
$$c_1x_1 + c_2x_2 + \ldots + c_nx_n \leq Z_0^B,$$

$$\ell_t = x_1 + x_2 + \ldots + x_n, \tag{13}$$

$$a_1y_1 + a_2y_2 + \ldots + a_ny_n \geq b,$$
$$c_1y_1 + c_2y_2 + \ldots + c_ny_n \leq Z_0^B,$$

$$\ell_1 = y_1 + y_2 + \ldots + y_n.$$

where $x_j, y_j \geq 0$, are the unknown variables. In this case, the y variables are used for the second LP.

4. Variable Freeing

4.1. Variable Freeing Theorem

If given n variables in an equality constraint form with all coefficients as integers as given in (14), then one of the n variables can be freed of integral restrictions.

In other words, one of the n variables is not supposed to be restricted to integer.

Variable freeing was first used in [39].

$$a_1x_1 + a_2x_2 + \ldots + a_jx_j + \ldots + a_{n-1}x_{n-1} + a_nx_n = b. \tag{14}$$

where a_j and b are integral constants and the unknown variables are all integers $\forall j = 1, 2, \ldots, n.$

If all the variables are integers, it is not necessary to restrict one of variables (x_j) out of n variables. This variable is a free variable and the integer problem becomes lighter in complexity.

Proof

From (14) the variable (x_j) can be made the subject as given in (15).

$$a_jx_j = b - (a_1x_1 + a_2x_2 + \ldots + a_{n-1}x_{n-1} + a_nx_n). \tag{15}$$

In this case, a sum of integers $(a_1x_1 + a_2x_2 + \ldots + a_{n-1}x_{n-1} + a_nx_n)$ in which one integer is being subtracted from another integer (b), then the difference must also be an integer.

According to constraint (15), the variable x_j is a free variable. There is no need to restrict this variable.

4.2. Strength of Variable Freeing in a Branch and Bound Setting

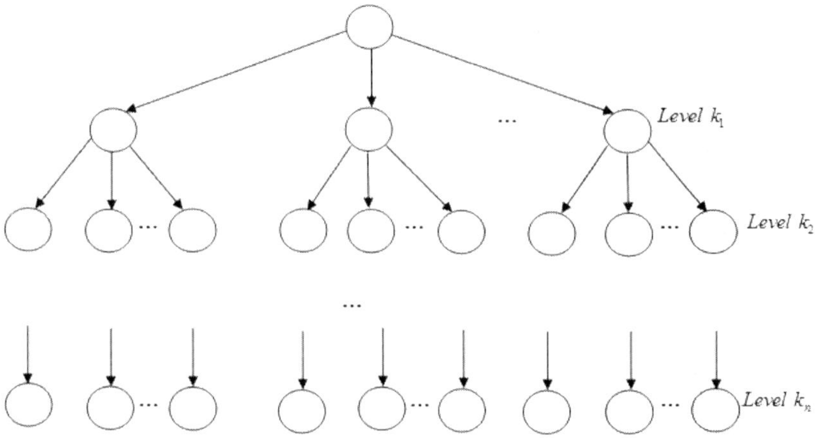

Figure 1. Nodes in a branch and bound setting.

In the worst case when all variables are restricted to integer:

At *Level* k_1 there are k_1 nodes.

At *Level* k_2 there are $k_1 k_2$ nodes.

...

At *Level* k_j there are $k_1 k_2 ... k_j$ nodes.

...

At *Level* k_n there are $k_1 k_2 ... k_j ... k_n$ nodes.

Freeing variable x_j means we are reducing the number of variables by a factor k_j.

In other words, the number of nodes reduces from $k_1 k_2 ... k_j ... k_n$ to $k_1 k_2 ... k_n$.

This shows how important freeing a variable is to a linear integer problem.

5. Reformulation of a Knapsack Problem

5.1. Reformulation by Introducing Binary Variables

From the expression in (11) binary variables can be introduced as given in (16).

$$x_1 + x_2 + \ldots + x_n = \ell_1 z_1 + \ell_2 z_2 + \ldots + \ell_t z_t. \tag{16}$$

where z_j are binary variables and (18) must hold.

$$z_1 + z_2 + \ldots + z_t = 1. \tag{17}$$

In other words, the knapsack problem in (4) can be formulated as given in (18).

Minimise $Z = c_1 x_1 + c_2 x_2 + \ldots + c_j x_j + \ldots + c_n x_n$,

Such that: $a_1 x_1 + a_2 x_2 + \ldots + a_j x_j + \ldots + a_n x_n \geq b$,

$$c_1 x_1 + c_2 x_2 + \ldots + c_n x_n \leq Z_0^B, \tag{18}$$

$$x_1 + x_2 + \ldots + x_n = \ell_1 z_1 + \ell_2 z_2 + \ldots + \ell_t z_t.$$

$$z_1 + z_2 + \ldots + z_t = 1.$$

where x_j is free and basic in the continuous optimal solution.

The reformulated knapsack problem given in (18) is easier to solve than the original form which is given in (4). This formulation has a weakness as the knapsack problem increases in size. When the number of variables becomes big, the bounding range $(\ell_t - \ell_1)$ also becomes big resulting in the number of binary variables becoming unmanageable.

5.2. Reformulation as a Bounded Variable

The challenge in Section 5.1 can not only be alleviated by avoiding binary variables but by formulating the problem as a bounded variable problem (19). The simplex method is efficient for such a problem.

Minimise $Z = c_1x_1 + c_2x_2 + ... + c_jx_j + ... + c_nx_n$,

Such that:

$$c_1x_1 + c_2x_2 + ... + c_jx_j + ... + c_nx_n \le Z_0^B, \tag{19}$$

$$x_1 + x_2 + ... + x_j + ... + x_n - x_\ell = 0.$$

where x_ℓ is an integer but bounded as $\ell_1 \le x_\ell \le \ell_t$.

In this form, the knapsack problem is neater with a constraint less than the form in (18).

6. Proposed Algorithm

The knapsack linear problem is solved using the following steps.

Step 1: Determine the objective bound Z_0^B.

Step 2: Use the objective bound to generate the variable sum limits ℓ_1 and ℓ_t.

Step 3: Solve the knapsack problem as a bounded variable problem.

6.1. Numerical Illustration

Minimize

$$Z = 162x_1 + 38x_2 + 26x_3 + 301x_4 + 87x_5 + 5x_6 + 137x_7,$$

Such that:

$$165x_1 + 45x_2 + 33x_3 + 279x_4 + 69x_5 + 6x_6 + 122x_7 \geq 18773,$$

where $x_j \geq 0$ and integer $\forall j$.

Step 1:

$$Z_0^B = \text{Min}[46818;15884;14794;20468;23751;15645;21098] = 14794$$

and x_3 is basic in the continuous optimal solution.

Step 2: Variable sum limits ℓ_1 and ℓ_t.

Maximize $\ell_t = x_1 + x_2 + x_3 + x_4 + x_5 + x_6 + x_7$,

Such that:

$$165x_1 + 45x_2 + 33x_3 + 279x_4 + 69x_5 + 6x_6 + 122x_7 \geq 18773;$$
$$162x_1 + 38x_2 + 26x_3 + 301x_4 + 87x_5 + 5x_6 + 137x_7 \leq 14794;$$

$\ell_t \leq 578.33 \therefore \ell_t = 578.$

Minimize $\ell_1 = x_1 + x_2 + x_3 + x_4 + x_5 + x_6 + x_7,$

Such that:

$$165x_1 + 45x_2 + 33x_3 + 279x_4 + 69x_5 + 6x_6 + 122x_7 \geq 18773;$$
$$162x_1 + 38x_2 + 26x_3 + 301x_4 + 87x_5 + 5x_6 + 137x_7 \leq 14794;$$

$\ell_1 \leq 568.43 \therefore \ell_1 = 569.$

In other words: $569 \leq \ell \leq 578.$

Step 3: Bounded variable problem

Minimize

$$Z = 162x_1 + 38x_2 + 26x_3 + 301x_4 + 87x_5 + 5x_6 + 137x_7,$$

Such that:

$$165x_1 + 45x_2 + 33x_3 + 279x_4 + 69x_5 + 6x_6 + 122x_7 \geq 18773;$$
$$x_1 + x_2 + x_4 + x_5 + x_6 + x_7 - x_\ell = 0.$$

where x_3 is now free and $569 \leq x_\ell \leq 578.$

Using the standard branch and bound algorithm within the proposed reformulation, it takes **35** sub-problems to verify optimality. Without any reformulation and solving the problem directly, it takes **1351** branch and bound sub-problems to verify optimality.

7. Computational Experiments

Over a hundred randomly generated knapsack problems were solved using the proposed reformulation. In all of these randomly generated knapsack problems, the number of branch and bound sub-problems required to verify optimality significantly decreased after the reformulation process. Visual Basic Software Version 10, was used for the computational experiments.

8. Limitations

This is a branch and bound related approach and there is no guarantee that the number of branches or sub-problems required to verify the optimal solution will not explode to unmanageable levels as the number of variables increase.

Conclusion

A lot has been done in terms of improving the branch and bound related algorithms. Cuts were incorporated within the branch and bound context to come up with what is now known as the branch and cut algorithm. Pricing was also incorporated within the branch and bound algorithm to produce the branch and price algorithm. Researchers went further to combine the three techniques, branch and bound, pricing and cutting planes to come up with the branch, cut and price. Even after all of these efforts, the knapsack problem is still in the NP Complete class of very difficult problems. Variable freeing is a new idea and there is hope that it may provide researchers with a new avenue for more research to solve this difficult problem.

The chapter provided a technique to free a single variable before solving the knapsack problem. Integral restriction is a nuisance when solving linear integer models. Variable freeing is a new idea in the history of integer programming.

FURTHER RESEARCH

Attempts will be made in the future to extend the proposed method to the general integer problem.

At the moment, there is no efficient and consistent general purpose method for the general integer problem.

REFERENCES

[1] Vasilchikov, V. 2018. On a Recursive-Parallel Algorithm for Solving the Knapsack Problem. *Automatic Control and Computer Sciences,* 52. 810 - 816. 10.3103/S014641161807026X.

[2] Bhattacharjee, K. K., Sarmah, S. P. 2014. Shuffled frog leaping algorithm and its application to 0/1 knapsack problem, *Applied Soft Computing,* Vol. 19. 252 - 263.

[3] Simon, J., Aruna Apte, A., Regnier, E. 2017. An application of the multiple knapsack problem: The self-sufficient marine, *European Journal of Operational Research,* Vol. 256 (31), 868 - 76.

[4] Lahyani, R., Chebil, K., Khemakhem, M., Coelho, L. C. 2019. Matheuristics for solving the Multiple Knapsack Problem with Setup, *Computers and Industrial Engineering Volume*, Vol. 129. 76 - 89.

[5] Martello, S., Monaci, M. 2020. Algorithmic approaches to the multiple knapsack assignment problem, *Omega,* Vol. 90.

[6] Bienstock, D., Faenza, Y., Malinović, I., Mastrolilli, M., Zuckerberg, M. 2021. On inequalities with bounded coefficients and pitch for the min knapsack polytope, *Discrete Optimization, Operations Research Letters,* Vol. 49 (3), 357 - 364.

[7] Fampa, M., Lubke, D., Wang, F., Wolkowicz, H. 2020. Parametric convex quadratic relaxation of the quadratic knapsack problem, *European Journal of Operational Research*, Vol. 281(16). 36 - 49.

[8] Dahmani, I., Hifi, M., Saadi, T., Yousef, L. 2020. A swarm optimization-based search algorithm for the quadratic knapsack problem with conflict Graphs, *Expert Systems with Applications,* Vol. 14815.

[9] Wu, Z., Jiang, B., Karimi, H. R. 2020. A logarithmic descent direction algorithm for the quadratic knapsack problem, *Applied Mathematics and Computation,* Vol. 36915.

[10] Fomeni, F. D., Kaparis, K., Letchford, A. N. 2020. A cut-and-branch algorithm for the Quadratic Knapsack Problem, *Discrete Optimization.*

[11] Wei, Z., Hao, J. K. 2019. Iterated two-phase local search for the Set-Union Knapsack Problem, *Future Generation Computer Systems,* Vol. 101. 1005 - 1017.

[12] Bandyopadhyay, S., Maulik, U., Chakraborty, R. 2013. Incorporating ϵ-dominance in AMOSA: Application to multiobjective 0/1 knapsack problem and clustering gene expression data, *Applied Soft Computing,* Vol. 13(5). 2405 - 2411.

[13] Zouache, D., Moussaoui, A., Abdelaziz, F. B. 2018. A cooperative swarm intelligence algorithm for multi-objective discrete optimization with application to the knapsack problem, *European Journal of Operational Research*, Vol. 264(11). 74 - 88.

[14] Abdel-Basset, M., El-Shahat, D., Faris, H., Mirjalili, S. 2019. A binary multi-verse optimizer for 0-1 multidimensional knapsack problems with application in interactive multimedia systems, *Computers and Industrial Engineering*, Vol. 132. 187 - 206.

[15] Arin, A., Rabadi, R. 2016. Local search versus Path Relinking in metaheuristics: Redesigning Meta-RaPS with application to the multidimensional knapsack problem, *Applied Soft Computing*, Vol. 46' 317 - 327.

[16] Wang, L., Yang, Y., Ni, H., Ye, We, Pardalos, P. M. 2015. A human learning optimization algorithm and its application to multi-

dimensional knapsack problems, *Applied Soft Computing*, Vol. 34. 736 - 743.

[17] Lai, X., Jin-Kao Hao, Fu, Z. H., Yue, D. 2020. Diversity-preserving quantum particle swarm optimization for the multidimensional knapsack problem, *Expert Systems with Applications*, Vol. 1491.

[18] Lai, X., Hao, J. K., Fu, Z. H., Yue, Y. 2020. Diversity-preserving quantum particle swarm optimization for the multidimensional knapsack problem, *Expert Systems with Applications*, Vol. 1491. Amiri, A. 2020. A Lagrangean based solution algorithm for the knapsack problem with setups, *Expert Systems with Applications,* Vol. 1431.

[19] Amiri, A. 2020. A Lagrangean based solution algorithm for the knapsack problem with setups, *Expert Systems with Applications,* Vol. 1431.

[20] El Yafrani, M., Ahiod, B. 2017. A local search based approach for solving the Travelling Thief Problem: The pros and cons, *Applied Soft Computing,* Vol. 52.

[21] Majid Khonji, M., Karapetyan, A., Elbassioni, K., Chau, S. C. K. 2020. Complex-demand scheduling problem with application in smart grid, *Theoretical Computer Science,* Vol. 76121. 34 - 50.

[22] Oprea, S. V., Bâra, A., Ifrim, G. A., Coroianu, L. 2019. Day-ahead electricity consumption optimization algorithms for smart homes, *Computers and Industrial Engineering*, Vol. 135. 382 - 401.

[23] Chakraborty, T., Misra, I. S. 2020. A novel three-phase target channel allocation scheme for multi-user Cognitive Radio Networks, *Computer Communications*, Vol. 15415. 18 - 39.

[24] Samavati, M., Essam, D., Nehring, M. Sarker, R. 2017. A methodology for the large-scale multi-period precedence-constrained knapsack problem: an application in the mining industry, *International Journal of Production Economics*, Vol. 193. 2 - 20.

[25] Alanne, A. 2004. Selection of renovation actions using multi-criteria "knapsack" model, *Automation in Construction*, Vol. 13. 377 - 391.

[26] Delgado-Antequera, L., Caballero, R., Sánchez-Oro, J., Colmenar, J. M., Rafael Martí, R. 2020. Iterated greedy with variable

neighborhood search for a multiobjective waste collection problem, *Expert Systems with Applications,* Vol. 1451.

[27] Hassanzadeh, A., Xu, Z., Stoleru, R., Gu, G. 2016. Michalis Polychronakis. 2016. PRIDE: A practical intrusion detection system for resource constrained wireless mesh networks, *Computers and Security,* Vol. 62. 114 - 132.

[28] Balas, E., Ceria, S., Cornuéjols, G., Natraj, N. (1996). Gomory cuts revisited. *Operations Research Letters*, 19:1 - 9.

[29] Alrabeeah, M., Kumar, S., Al-Hasani A., Munapo, E. and Eberhard, A. 2019. Computational Enhancement in the Application of the Branch and Bound Method for Linear Integer Programs and Related Models, *International Journal of Mathematical, Engineering and Management Sciences,* Vol. 4, No. 5, 1140 - 1153, 2019, https://dx.doi.org/10.33889/IJMEMS.2019.4.5-090.

[30] Brunetta, L., Conforti, M., Rinaldi, G. 1997. A branch and cut algorithm for the equicut problem. *Mathematical Programming,* 78, 243 - 263.

[31] Mitchell, J. E. 2001. Branch and cut algorithms for integer programming: In Floudas, C. A. and Pardalos, P. M. (Eds.), *Encyclopedia of Optimization*, Kluwer Academic Publishers.

[32] Padberg, M., Rinaldi, G. 1991. A branch and cut algorithm for the resolution of large-scale symmetric traveling salesman problems. *SIAM Review,* 33(1), 60 - 100.

[33] Barnhart, C., Johnson, E. L., Nemhauser, G. L., Savelsbergh, M. W. P., Vance, P. H. 1998. Branch and price column generation for solving huge integer programs. *Operations Research,* 46, 316 - 329.

[34] Salvelsbergh, M. W. P. 1997. A branch and price algorithm to solve the generalized assignment problem. *Operations Research,* 45, 381 - 841.

[35] Barnhart, C., Hane, C. A., Vance, P. H. 2000. Using branch-and-price-and-cut to solve origin-destination integer multicommodity flow problems. *Operations Research,* 48, 318 - 326.

[36] Fukasawa, R., Longo, H., Lysgaard, J., Poggi de Aragao, M., Uchoa, E., Werneck, R. F. 2006. Robust branch-and-cut-price for the

Capacitated vehicle routing problem. *Mathematical Programming,* Series A 106, 491 - 511.

[37] Ladányi, L., Ralphs, T. K., Trotter, L. E. (Jr.). 2001. Branch, cut and Price: Sequential and Parallel, in *Computational Combinatorial Optimization,* Naddef, N. and Jüenger, M., eds, Springer, Berlin, 223.

[38] Munapo, E. 2020. Improving the Optimality Verification and the Parallel Processing of the General Knapsack Linear Integer Problem, *Research Advancements in Smart Technology, Optimization, and Renewable Energy*, Chapter 3, IGI Global, pp. 37 - 52.

[39] Munapo, E. 2021. Branch Cut and Free Algorithm for the General Linear Integer Problem. In: Vasant P., Zelinka I., Weber G. W. (eds). Intelligent Computing and Optimization. ICO 2020. *Advances in Intelligent Systems and Computing*, vol 1324. Springer, Cham. https://doi.org/10.1007/978-3-030- 68154-8_44.

About the Editors

Pandian Vasant is an Editor-in-Chief of International Journal of Energy Optimization and Engineering (IJEOE) and member of MERLIN research group of Ton Duc Thang University. He holds PhD in Computational Intelligence (UNEM, Costa Rica), MSc (University Malaysia Sabah, Malaysia, Engineering Mathematics) and BSc (Hons, Second Class Upper) in Mathematics (University of Malaya, Malaysia). His research interests include Soft Computing, Hybrid Optimization, Innovative Computing and Applications. He has co-authored research articles in journals, conference proceedings, presentation, special issues guest editor, book chapters (290 publications indexed in ResearchGate) and General Chair of EAI International Conference on Computer Science and Engineering in Penang, Malaysia (2016) and Bangkok, Thailand (2018). In the year 2009 and 2015, Dr. Pandian Vasant was awarded top reviewer and outstanding reviewer for the journal Applied Soft Computing (Elsevier) respectively. He has 31 years of working experiences at the various universities. Currently he is an Editor-in-Chief of International Journal of Energy Optimization & Engineering, and Member of AMS (USA), NAVY Research Group (TUO, Czech Republic) and MERLIN Research Group (TDTU, Vietnam). H-Index Google Scholar = 33; i-10-index = 107.

Valeriy Kharchenko: Date of birth 18 December 1938 Nationality Russian Key qualifications: researcher and consultant in a wide range of energy saving fields, including energy policy at national and international level, energy RTD strategy, RUE and RES. Education 1994 - 1995 Tacis Training Programme (Manager in energy) 1994 Tashkent State Technical University, Uzbekistan (professor) 1987 State Higher Training Courses on Patenting and Invention (Certificate with distinction) 1987 Institute of Electronics, Moscow (Doctor of Engineering Sciences Degree) 1969 Institute of Inorganic Chemistry, Novosibirsk (Candidate of Chemistry Degree) 1956 - 1961 Tashkent Polytechnic Institute (Technical University) (Technology engineer) International activity experience: 1994-1996 – counterpart expert in the Project of EU "Monitoring and evaluation of the energy (non nuclear) projects implementation in Uzbekistan" (carried out by European companies); 1998-2000 - the expert for evaluation of submitted proposals INTAS (ID 716); 1996-2000 –senior expert of the Central Asia Energy Advisory Group (a project under the EC DG XVII's SYNERGY Programme) with obligations included duties for formation of expert groups, peering and estimation of their reports and preparing general Interim and Final annual reports; 2000-2001 years -UNDP Program, Global Environmental Facility, Government of Uzbekistan (Joint Project UZB/98/G42/A/1G/99), expert at the premises. Among of obligations - evaluation and reception of reports of local expert groups on various aspects of the organization of heating and hot water supply system of Tashkent city Area of Scientific activity. Development of new technological approaches for superpure silicon fabrication, improvement of silicon epitaxial layers technology, application of silicon layers in various devices, including PVcells, application of installations on the basis of solar cells in practice, development of new approaches for production of silicon for solar cells, application of a solar energy for electricity and heat production; Participation in a number of international projects of the Eurocommission and UNDP on an energy policy, energy saving (RUE) and renewable energy sources (RES), works on increase of efficiency of use of energy in systems of heating and hot water supply. He currently holds the position of Chief Scientific Officer of the Federal Scientific

Agroengineering Center. He is the official reviewer of a number of international journals, including Solat Energy, Deputy Editor-in-Chief of European Journal of Renewable Energy ", member of the editorial boards of several international journals, honorary professor of the Institute of Hydropower and Renewable Energy of the National Research University" MEI. Under his edition a number of books were published by IGI Global (USA).

Vladimir Panchenko: From 2003 to 2009 studied at "Bauman Moscow State Technical University" at the faculty "Power Engineering " with qualification of engineer. From 2009 to 2012 studied at the graduate school of the "All-Russian Scientific Research Institute of Agriculture Electrification", specializing in "Renewable Energy Installations". Passed an internship program on the International Energy Program in the Solar Energy Center (SEC), the Certificate under the Indian Technical and Economic Cooperation (ITEC), the Program of the Ministry of External Affairs (MEA), the Government of India - India, Delhi, November-December 2011. Participated in the International Conference-Exhibition Intersolar Europe 2012, Germany, Munich, June 11 - 15, 2012 as part of the delegation of the Ministry of Energy of the Russian Federation. Ph.D. thesis "Development and research of solar photovoltaic thermal module with a paraboloid-type concentrator" was protected on July 2, 2013. From 2013 to 2017 head of the laboratory "Solar photovoltaic modules with an experimental - technological site". The Council of Young Scientists of the VIESH was elected the Chairman of the Council of Young Scientists and was appointed a member of the Academic Council of the VIESH in 2014 - 2017. In 2014 was appointed as a contact person and a member of the International Solar Energy Society of the Russian section of the community. Since September 2015 associate professor at the "Russian University of Transport". Since 2017, a senior researcher at the "Solar Energy Laboratory" of "Federal Scientific Agroengineering Centcr VIM". Passed the advanced training in 2017 in "Russian University of Transport" the program "Using e-learning and distance educational technologies in the implementation of educational programs on distance learning". In 2016 awarded the grant "Young lecturer of Russian University of Transport"'. In 2018 awarded the Scholarship of the President of the Russian Federation for 2018 - 2020 for young scientists and graduate students engaged in advanced research and development in priority areas of the modernization of the Russian economy. Has 3 medals of exhibitions, 3 diplomas of the winner of competitions, participated in more than 80 conferences and exhibitions of various levels, has more than 120 publications, including patents and educational publications. Area of interest includes three-dimensional modeling, use of solar energy and electrical transport.

J. Joshua Thomas is a senior lecturer at KDU Penang University College, Malaysia since 2008. He obtained his PhD (Intelligent Systems Techniques) in 2015 from University Sains Malaysia, Penang, and Master's degree in 1999 from Madurai Kamaraj University, India. From July to September 2005, he worked as a research assistant at the Artificial Intelligence Lab in University Sains Malaysia. From March 2008 to March 2010, he worked as a research associate at the same University. Currently, he is working with Machine Learning, Big Data, Data Analytics, Deep Learning, specially targeting on Convolutional Neural Networks (CNN) and Bi-directional Recurrent Neural Networks (RNN) for image tagging with embedded natural language processing, End to end steering learning systems and GAN. His work involves experimental research with software prototypes and mathematical modelling and design He is an editorial board member for the Journal of Energy Optimization and Engineering (IJEOE), and invited guest editor for Journal of Visual Languages Communication (JVLC-Elsevier). He has published more than 30 papers in leading international conference proceedings and peer reviewed journals.

G.-W. Weber is a Professor at Poznan University of Technology, Poznan, Poland, at Faculty of Engineering Management, Chair of Marketing and Economic Engineering. His research is on OR, financial mathematics, optimization and control, neuro- and bio-sciences, data mining, education and development; he is involved in the organization of scientific life internationally. He received his Diploma and Doctorate in mathematics, and economics / business administration, at RWTH Aachen, and his Habilitation at TU Darmstadt. He held Professorships by proxy at University of Cologne, and TU Chemnitz, Germany. At IAM, METU, Ankara, Turkey, he was a Professor in the programs of Financial Mathematics and Scientific Computing, and Assistant to the Director, and he has been a member of further graduate schools, institutes and departments of METU. Further, he has affiliations at the universities of Siegen, Ballarat, Aveiro, North Sumatra, and Malaysia University of Technology, and he is "Advisor to EURO Conferences".

INDEX

A

B

C

D

E

F

G

H

I

S

T

V

W